Design, Development and Sensemaking of Human-Robot Interaction in Care Settings

Felix Carros

Design, Development and Sensemaking of Human-Robot Interaction in Care Settings

Springer Vieweg

Felix Carros
Universität Siegen
Siegen, Germany

Department for Information Systems and New Media, Faculty III, University of Siegen, Germany
Submitted in Fulfillment of the Requirements for the PhD Degree: Doktor rerum politicarum (Dr. rer. pol.)
1. Reviewer: Volker Wulf (Department for Information Systems and New Media, University of Siegen)
2. Reviewer: Claudia Müller (Chair of Information Systems, esp. IT for the Aging Society, University of Siegen)
Supervisors: Volker Wulf and Claudia Müller
Date of Submission: January 20, 2023

ISBN 978-3-658-45232-2 ISBN 978-3-658-45233-9 (eBook)
https://doi.org/10.1007/978-3-658-45233-9

This Springer Vieweg imprint is published by the registered company Springer Fachmedien Wiesbaden GmbH, part of Springer Nature.
The registered company address is: Abraham-Lincoln-Str. 46, 65189 Wiesbaden, Germany

Acknowledgements

This document is my dissertation titled "Design, Development and Sensemaking of Human-Robot Interaction in Care Settings," and describes my research activities between 2018 and 2022 at the Institute for Information Systems and New Media, University of Siegen. It was influenced by my involvement as a student in the research cooperation between the university and a care home in the period from 2015 to 2018.

Although my name appears on the front page of this document, it could not have been accomplished without the cooperation and assistance of others. I am grateful to everyone who helped me reach this point. I believe this is a shared achievement. I received assistance in various areas for an extended period, and it is impossible to name everyone who helped me along the way because there were so many of you. As a result, I will refrain from using names and hope that each of you understands that I am referring to you. This support extended beyond immediate assistance with my research; it included meaningful conversations, feelings of connection, and moments of discovery. I'm sure I received assistance that I was not even aware of, people opening doors or helping me to overcome obstacles I was not aware existed. The assistance I received allowed me to create this work, and it would not have been possible without you.

I would like to express my gratitude to all the participants and institutions who granted me access to their resources. Special thanks to those individuals I have pseudonymized, you were critical to the success of this project. I understand that I have made significant demands at times, and I appreciate the effort you have put in to make our work possible. Your insights have been invaluable, and I am honored to have been a part of your lives. I acknowledge that this is not the norm.

I recognize that this was only achievable due to your efforts. You made me see, and I am grateful for that.

Although I am profoundly thankful, I think I also need to express regret. This thesis has consumed a significant amount of my time and energy. As a result, I might have neglected personal relationships and missed important events, conversations, and visits. I deeply appreciate the patience and understanding shown by my friends and family during this period and know that I was not the only one that has been investing time and nerves to create this work. Thank you for enabling this work.

Abstract

Western societies' demographic shift toward an older population has been a long-term trend. Consequently, the number of people in need of care services has increased. The number of people employed in the care sector is not increasing at the same rate. As a result, the existing workforce and the system are under strain. At the same time, technology has advanced rapidly. Social robots are one of these innovations that are gradually becoming a part of everyday life. Social robots' new functional and interactive capabilities make them relevant in care settings.

This dissertation investigates the potential of a social robot in care homes to assist the workforce and contribute to the well-being of residents. Several studies are presented that use a socio-informatics approach to make sense of social robots in care homes. They investigate how residents and caregivers interact with social robots, analyze and shape their appropriation, and how informed decisions can be made for the development of the software and hardware of the robots. The findings indicate that social robots can play a role in care homes. However, they can only assist and not replace care workers. They broaden the possibilities for social care workers and, when integrated into daily work practices, can provide small breaks. They can also have a positive impact on residents by motivating them to engage in cognitive and physical training.

We chose to explore the field using a participatory approach to design the robot's functionalities. Research and development was done within care homes together with the people working and living there. The findings aim to contribute to the HCI and HRI community by demonstrating the benefits, challenges, and ambiguities of long-term use of social robots in care homes.

List of Publications

This dissertation consists of an accumulation of publications that have been accepted and published. Chapter 4 corresponds to P1, Chap. 5 to P2, Chap. 6 to P3, Chap. 7 to P4, Chap. 8 to P5, and Chap. 9 to P6. The list of all publications is in Table 1 below. The publications are a shared result. I thank all my co-authors for the collaboration on these.

Table 1 List of publications that are part of this dissertation

No.	Publication Details	Outlet	Status
P1	Carros, F., Meurer, J., Löffler, D., Unbehaun, D., Matthies, S., Koch, I., Wieching, R., Randall, D., Hassenzahl, M., Wulf, V. 2020. *Exploring Human-Robot Interaction with the Elderly: Results from a Ten-Week Case Study in a Care Home* Proceedings of the 2020 CHI Conference on Human Factors in Computing Systems. (April, 2020). https://doi.org/10.1145/3313831.3376402 [Reference in bibliography: [69]]	Conference Proceedings	Published
P2	Carros, F., Schwaninger, I., Preussner, A., Randall, D., Wieching, R., Fitzpatrick, G., Wulf, V. 2022. *Care Workers Making Use of Robots: Results of a Three-Month Study on Human-Robot Interaction within a Care Home.* Proceedings of the 2022 CHI Conference on Human Factors in Computing System. (May, 2022). https://doi.org/10.1145/3491102.3517435 [Reference in bibliography: [71]]	Conference Proceedings	Published

(continued)

Table 1 (continued)

No.	Publication Details	Outlet	Status
P3	Schwaninger, I., Carros, F., Weiss, A., Wulf, V. and Fitzpatrick, G. (2022). *Video connecting families and social robots: from ideas to practices putting technology to work* Proceedings of the Universal Access in the Information Society. https://doi.org/10.1007/s10209-022-00901-y [Reference in bibliography: [276]]	Journal	Published
P4	Carros, F., Langendorf, J., Randall, D., Wieching, R., Wulf, V. (2022). *Citizen Participation in Social Robotics Research* Published in the Book "Meaningful Futures with Robots—Designing a New Coexistence" [Reference in bibliography: [68]]	Book Chapter	Published
P5	Störzinger, T., Carros, F., Wierling, A., Misselhorn, C., Wieching, R. (2020) *Categorizing Social Robots with Respect to Dimensions Relevant to Ethical, Social and Legal Implications* Proceedings of the i-com Journal, Vol. 19 (Issue 1), pp. 47–57. https://doi.org/10.1515/icom-2020-0005 [Reference in bibliography: [295]]	Journal	Published
P6	Carros, F., Störzinger, T., Wierling, A., Preussner, A., Tolmie, P. (2022). *Ethical, Legal & Participatory Concerns in the Development of Human-Robot Interaction—Lessons from eight research projects with social robots in real-world scenarios.* Proceedings of the i-com Journal, Vol. 21 (Issue 2), pp. 299–309. https://doi.org/10.1515/icom-2022-0025 [Reference in bibliography: [73]]	Journal	Published

Contents

Part II Publications

Abbreviations

AI	Artificial Intelligence
COP	Community of Practice
COVID-19	Coronavirus Disease 2019
CSCW	Computer-Supported Cooperative Work
ELSI	Ethical, Legal and Social Implications
EU	European Union
EUD	End-User Development
GDPR	General Data Protection Regulation
HCI	Human–Computer Interaction
HRI	Human–Robot Interaction
HTML	HyperText Markup Language
ICT	Information and Communication Technology
IS	Information Science
IT	Information Technology
MEESTAR	Model for the Ethical Evaluation of Socio-Technological Arrangements
NGO	Non-Governmental Organization
NRW	North Rhine-Westphalia
OHRI	Older Human–Robot Interaction
PD	Participatory Design
RRI	Responsible Research and Innovation

List of Figures

List of Tables

Part I
Foundations

The first section of the dissertation presents its foundations and context. Chapter 1, the introduction, explains the subject and my own motivation, as well as its limitations. Chapter 2 places the dissertation's work in the context of related work and demonstrates the need for change within the sector, and explains why social robots can contribute to this change. In Chap. 4, the last chapter of this first part I present the methodology of this work, how I approached the field and learned from it and reflect on the experiences in the field.

Introduction

1

1.1 Problem Definition

Today we can say with certainty that the care system in Western societies, and specifically in Germany, is under pressure. In the coming decade this pressure is likely to increase, not decrease. The reasons for this trend are manifold, but the main driving force of this development is demographic change. German society is growing older and has fewer younger people. The number of people in Germany older than 80 years is predicted to grow from 6.2 million in 2022 to 8.9–10.5 million in 2050. By comparison, in 1970 Germany had 1.15 million people over 80 years old [54, 117]. Another reason is that the number of people working in care is not growing at the same pace; in fact, many positions remain unfilled [19]. In summary, the number of people in need of care grows while the number of people working in care does not, resulting in a shortage of workers and a need for assistance. Far from being a surprising development, this has been projected since decades past, but today we have a clearer picture of it.

This raises a question: How can care work be assisted, and can technology play a role in this assistance? In this work, we look at social robots and how they can (or cannot) be of assistance in the care field. Social robots in care settings are not new; they existed long before this work was written and appear in a range of topics including nutrition, entertainment, loneliness or hygiene. The global research community has worked intensively on social robots in care settings as it is a common vision to help the care sector with robots. The general idea is to assist the care sector with robotic systems in order to relieve the work force, something that previously was done in the industrial sector [112]. In the industrial sector, it is nowadays not uncommon to see a variety of robots working together to create a product, sometimes fully automated. Human workers do not have the same jobs as they used to in places

F. Carros, *Design, Development and Sensemaking of Human-Robot Interaction in Care Settings*, https://doi.org/10.1007/978-3-658-45233-9_1

that heavily use robots for automation. Many of the traditional jobs vanished in these places, and workers are working around the robots to accommodate or control them. But there is a fundamental difference between the industrial sector and the care sector. Industries are working with commodities, trying to create something more valuable out of them, whereas care settings are working with humans and do not try to create monetary value. Although it may be ethically acceptable for a society to use robots in factories, this cannot be transferred to other areas, for each area must be evaluated according to societal values and needs. The care sector is a sensitive field with vulnerable people, so even if, technology-wise, the robots can become a solution, it remains unclear whether that is a desirable outcome. Also, apart from the ethical point of view, there is another major difference, namely the complexity of robotic systems working with humans. Working with commodities is to some degree predictable, whereas working with people is less so.

Humans are difficult to predict and do not behave rationally. Social robots were constructed to interact with humans, which makes their designated task more difficult. Most industrial robots have only one task or several predefined tasks, such as screwing a screw into an automotive part. Social robots, on the other hand, cannot predict what will happen next; they do not have a predefined schedule, and they must interact with humans, who may change their mind or act spontaneously. Therefore, it is different to program these robot because instead of following a predefined schedule, they must react to all kind of behaviors, like residents asking questions or humans walking in the path of the robots. These reactions are not yet fully working: robots can react only to certain behaviors, and most of the time their coding does not tell them how to react accordingly. This is no surprise, as social robots are still fairly new and most have been developed within controlled environments in labs.

As social robots are increasingly being developed and the situation in the care sectors persists, there is a need for applied research on social robots in care homes. Only within the applied field is it possible to see how the technology appeals to the sector. Developing robot applications based on participatory design with all involved stakeholders and the evaluation of such developments in long-term studies within the care homes appears important to understand how such technology is working and what effects it has on the different stakeholder groups. Based on this knowledge, it is possible to conduct a societal discussion as to whether such technology in the field is a desirable outcome. Starting a discussion before this field work has begun appears unsuitable, for it is unclear how the use of social robots in care homes will unfold. In addition to this, the perception of what robots can do is highly shaped by fiction, which illustrates robots as machines with superpowers. However, this fiction is far from the truth, as social robots have many limitations. Only by working in the field with the practitioners and the technology is it possible to develop realistic applications for robots that are helpful for the field and within the limits of technological feasibility.

1.2 Motivation for this Work

The motivation for this work followed a vision that social robots can be a valuable contribution for care home residents. This vision came from my first encounters with social robots and the effects they had on me. At the time, I thought that my positive emotional reaction to the social robot Pepper would be easily translatable to care home residents. In addition, before beginning this dissertation I was already working on a research project in a care home. For more than two years, we tested how care home residents use tablets and health platforms. I learned a lot about the use of technology in this group. In that respect, this work also derives from practical motivations. At the beginning, there was the interest to do something good with my abilities and the technology and to improve life in care homes for the residents. In retrospect, this was a naive position, as it was too technocratic. I thought at that time that technology could work without the human factor, that the technology itself would not need to be taken care of. I thought that social robots would enter the care homes and that their presence and functionality would lead to a relief of the care workers. That notion has changed throughout my years of field research. I came to realize that care homes and the activities carried out their function in a complex interwoven social structure; merely dumping technology there does not lead to any significant change. In my experience, it instead results in a lot of money being spent on technology that ultimately remains unused.

I quickly learned that the integration of technology into this sensitive field requires a multidimensional, interdisciplinary approach in the sense that the research involves different disciplines working on the same topic, albeit with different aspects in mind. I can provide only a socio-informatics perspective. Despite this limitation, this work tried to be in exchange with other disciplines and adapted some of their viewpoints. Different disciplines and stakeholders with their specific opinions and experiences play a major role in this dissertation, such as engineering, computer science, ethics and legal.

In summary, the motivation to create something good with technology has remained with me, but I had to acknowledge that the task is more complex than I initially thought. Therefore, it is difficult for me to say that I have reached my goal. One robot is, as I'm writing this, used by the care home we used to work with. They requested to have it, so the care workers can use it in their daily work with the residents. This was part of my initial vision, but I have to admit I thought it would be different when I started. However, the path of this work has involved many people who have influenced me, and whom I have influenced. In the end, it has to be said that a single person or research group cannot have a significant influence on the use of technology in care because the field of application is too large and historically grown. However, insights can be shared with academic papers that provide insight to those reading them-insights that influence others in the development of technology. In this respect, the motivation of this work is to create a knowledge gain for the field but also for myself-not only for academic discourse, but also to bring this topic to the attention of mainstream society.

1.3 Main Contribution of the Dissertation

This dissertation offers insights in how care homes are using social robots in their daily lives. It offers three levels of insight: First, it shows the reader how care home residents interact with a social robot over a long term. We used the robots from 2017 until today (2022, ongoing) and have gained several insights on how the system is used in practice and how stakeholders (mainly residents and care workers of a care home) react to it. Second, it gives insights on the appropriation of a social robot in the daily working routines of care workers. We wanted to know how they are using such systems and how they integrated them into their ongoing work when they are completely free to use it. The goal was to learn how a sustainable integration of the social robot can work, in order to make it a useful tool for the people working in a care home. Third, we looked at ethical and social factors of using social robots in care homes. Because the setting is a vulnerable place that involves many different stakeholders, it is important to gain a holistic view to understand different disciplines working in that field. As this dissertation is adopting a practice-based approach, working in the field with a robotic system that is already advanced in the development, it was important to understand the complex environment in order to make a successful implementation.

This work aims at giving the reader insights on how social robots can be used in care homes, by exploring the long-term effects of such. Methods like Participatory Design, PraxLabs and Design Case Studies have yielded results that contribute to the discussion of whether social robots are beneficial for the care sector. We could show that the social robots will not solve the challenges but that they can be of assistance and that they are able to enhance the wellbeing of residents.

1.4 Structure of Thesis

The dissertation is organized in three sections: the foundation of the results, the publications, and the discussion and limitations of the results.

This first section (Chap. 1–3) introduces the background and setting of the dissertation. In Chap. 1 in the introduction. The context of this work is presented in Chap. 2 along with comparisons to similar literature. It demonstrates the need for change in the industry and the potential role social robots could play in it. The methodology of this work, including how I approached the field, learnt from it, and considered the experiences in the field, is presented in Chap. 3, the final chapter of this first section.

The publications are included in the second section of this dissertation. The core of this work comprises six publications, which I present. They are arranged in six Chaps.: 4 to 9. By emphasizing practice-based research with social robots and their use in care situations, the papers are thematically related.

The interaction between care home inhabitants and a social robot is the subject of Chap. 4. We considered utilization over the long run, how it affected the locals, and what personnel were required. In Chap. 5, we examined the care providers and how they used the robot in their day-to-day job. After years of building the robot, we wanted to investigate its potential for sustainable use. We aimed to integrate the robot into care workers' routines in order to achieve sustainable technology usage. The study we conducted in two care homes during the early stages of COVID-19 is discussed in Chap. 6, which focuses on how the residents and caregivers at care homes used technology in this initial stage, how they had to collaborate, and how they saw potential uses for robots. In Chap. 7, we examine how citizens can be involved in social robot research in healthcare settings. Care homes are a very delicate subject that calls for public discussion. We think that technological advancements in the healthcare industry should be explored and planned with citizens. The ethics of social robots are discussed in Chap. 8; autonomy, emotionality, sociality, and the effect on social skills are examined. I share the lessons I've learnt from working on several social robot research projects for three years in the final chapter (Chap. 9) of this section.

I review the findings from the earlier chapters in this dissertation's third section and come to conclusions and suggestions for the HCI community. Chapter 10 discusses the findings, compares them to those of other studies, and includes author views on the subject. I identify the limits of this dissertation in Chap. 11, which are primarily attributable to the qualitative practice-based research. Finally, in Chap. 12 I provide the conclusions I drew from the material that has been presented, as well as my viewpoint on the subject and ideas for additional research.

Related Work and Context 2

In this chapter, I write about related work and show how robots have developed over the years as well as how social robots are used in the context of care. I then proceed to explain the context of this dissertation and show the state of affairs in the demographic change and the status quo in care.

2.1 Related Work

2.1.1 Evolution of Robotics

Ideas about machines that work for humans date back at least as far as ancient Greek mythology [152], but only with modern advances in electronics has it become possible for such machines to become functioning robots capable of executing complex tasks.

One of the first mentions of the word "robot" can be traced back to the Czech word *robota*, which can be translated as "forced labor," "serf," or "laborer" [144, 240]. In 1917, the Czech author Joseph Capek wrote a story that describes *automats*; later, in 1921, his brother Karel Capek wrote a play called *Rossum's Universal Robots*. The play was a critique of the rapid development of technology and described robots as machines with growing capabilities that would at one point revolt against humans [63, 65, 144]. These two incidences are seen as the moment when the concept of robots entered popular consciousness [144]. In 1935, after the concept of robots had been used as a broader term for mechanised machines Karel Capek commented on the interpretation of the word "robot", he stated that he initially thought that robots are a product of biological chemistry, some sort of living cell that imitated life, "something like another alternative to life". He strongly refused

F. Carros, *Design, Development and Sensemaking of Human-Robot Interaction in Care Settings*, https://doi.org/10.1007/978-3-658-45233-9_2

the concept that mechanical machines could imitate life and saw this as an "unforgivable overvaluation of mechanics or as a severe insult to life." [64] (p. 265 f.) [5].

There are various definitions for the word "robot." The Robot Institute of America defined it as *"a reprogrammable, multifunctional manipulator designed to move materials, parts, tools, or specialized devices through various programmed motions for the performance of a variety of tasks"* [13, 185]. Oxford sees it as *"A machine capable of automatically carrying out a complex series of movements, esp. one which is programmable,"* [240] and Cambridge describes it as *"a machine controlled by a computer that is used to perform jobs automatically"* [62]. All descriptions contain the words "automation" or "programmed," insinuating these characteristics as important parts of robots. Robots should do things automatically, but the spectrum of what is done automatically has changed over the years with evolving technological possibilities.

Zamalloa et al. [349] see five generations of robots. The first is the "Pre-Robots" generation that consists mainly of pneumatic and hydraulic robots and that was predominant from the first design of a robot by Leonardo Da Vinci [116] until 1950. Pre-Robots were mainly automated machines used during industrialization put in motion by pneumatic or hydraulic systems. The second generation is called "First Manipulators," spanning 1950 to 1967. These were numerically controlled machines that were combined with the manipulators, such as mechanical arms. Both elements together created early forms of robots. They were used mainly within the automotive industry, often to grasp and place objects. The third generation were the "Sensorized Robots" from 1968 to 1977. In that time, robots became increasingly aware of their surroundings through the use of sensors, such as tactile sensors or cameras, which made different tasks possible for the robots. These robots were still used mainly in the industrial sector, and again predominantly in automotive manufacturing, but because of the new sensors, this third generation of robots were sometimes able to learn by demonstration and could respond to more than one challenge. A famous example was the Famulus robot released in 1973 by the company KUKA; it was an industrial robot with six electromechanical axes [285]. The fourth generation were "Industrial Robots" (1978–1999), identifiable by two specific characteristics: they had built-in computers and were reprogrammable. Additionally, around this time, some programming languages emerged that were dedicated to the programming of robots. Some of these Industrial Robots also started to employ artificial vision. Zamalloa et al. [349] describe the fifth and final generation as "Intelligent Robots," predominant between 2000 and 2017, during which time robots gained many abilities; as such, each model differed from the others. The Intelligent Robots can be categorized by a few characteristics: (1) advanced computing capabilities, making it possible to work in complex situations; (2) the ability to learn and act with a certain degree of logic with elements of artificial intelligence; (3) sophisticated sensors from which the information is computed within the robot, making it possible to act on several solid sources of information; and (4) collaborative elements, making it possible to work with humans or other robots.

In recent years, we have also seen an increase in the development of social robots [142, 264]. As their name suggests, social robots are not built for physical work, but rather for social interaction. They can nowadays be seen occasionally in public spaces (e.g., shopping centers, conferences), and more and more people are coming into contact with them. Social robots are different from industrial robots and are instead built to interact with people [50]. These robots are not primarily built to do physical work, but rather to speak or to foster an emotional response. Many of these robots look human- or animal-like; they usually do not look like strong machines. Indeed, they often have elements that make them look innocent (e.g., big childlike eyes) in order to evoke an emotional response and to dispel fear. Their purposes are manifold but can mainly be associated with the service industry: They can talk with people, touch them, or offer them some sort of companionship. The most famous of the social robots is probably NAO from Aldebaran Robotics in Paris (see Fig. 2.1). This model was used in many Human-Robot Interaction (HRI) studies and is probably the most studied social robot in the HRI community [32]. The Aldebaran company also produced the robot Pepper [243], which is taller and heavier than NAO. Pepper was used in most of the studies presented in this thesis.

There are many different categories of social robots. As one robot can be used for multiple tasks, the categories will be presented based on the application areas. According to Bartneck et al. [32] social robots can be differentiated into nine categories: (1) Service robots that mainly talk to people (robots that can serve as tour guides, receptionists, or sales promoters). (2) Robots that help with learning (these systems do not need a specific shape; only an advanced chatbot and software that enables learning). (3) Entertainment robots (includes subcategories like toy robots, pet robots, robots that perform in exhibitions and art installations, or even sex robots). (4) Robots in healthcare (this is the category of robots used in the studies of this thesis; robots in healthcare can be used in the context of elderly people in need of care, people with disabilities, as well as hospital and rehabilitation patients). (5) Personal assistant robots (these robots are mainly about giving information to the user; they often help in daily life tasks, like obtaining news or reporting the weather). (6) Service robots, which help with simple tasks, such as by delivering things, cleaning, or even acting as security guards by escorting people through dark parking lots to their cars (the social element is less strong in these applications but still exists as the robots are noticing the human and reacting to them). (7) Collaborative robots are mainly used in the industrial context. They are designed to work together with humans in industry applications (in contrast to the classic industrial robots that used to work without human counterparts). (8) Self-driving cars (robotic systems designed to communicate with the passenger but also with other cars and people on the road). (9) Remotely operated robots (there are many application areas for these; the most prominent are certainly those used in space and on other planets, but we also see more common developments like social robots that are remotely operated to welcome guests in hotels).

Fig. 2.1 NAO Robot. (Photo: Jérémy Barande [157])

As we see in the categories above, many of the applications have a practical background, and the systems have increasingly become part of people's daily lives. Through the penetration of public spaces, the term "robot" becomes more and more a reality and not something that people know from fiction; they now know it through real interactions. These interactions change how people think about robots, and they are shrinking the gap between what people think robots should be able to do (often shaped by fiction) and what robots are actually able to do.

The new capabilities of social robots make real-world scenarios possible. In order to create useful human-robot interactions, it seems crucial to test and develop these devices in practice, as people and robots tend to act and react differently in real-world settings than in controlled laboratory situations.

2.1.2 Social Robots in Care Settings

The existing research on social robots in care settings has covered a wide range of topics and is generally labeled as Human-Robot Interaction (HRI). Some researchers even created a term for this specific line of research: Older Human-Robot Interaction (OHRI) [321]. The state of the art includes care workers' perceptions of robots and the standards for their acceptability [213], and in another line of research the acceptance of older adults to robots [25, 154, 247]. Other researchers have looked into the ethics of using robots in care settings (e.g., [158, 319]); even the European Parliament looked into this topic and pointed out that robots in such settings should follow values like transparency, autonomy, dignity, equality and privacy [245]. Research has further looked into special domain contexts such as dementia care (see, e.g., [143, 151, 214]), rehabilitation [104, 175, 176] and disabilities [17, 128, 140]. Other researchers have looked into the status of robots as assistants and what possibilities (as well as limitations) they create [34, 224]. Although these studies report some benefits, as the findings of Bemelmans et al. [34] suggest, definitive conclusions are elusive due to limitations in the methodology, as many of the results are created under artificial conditions. Although this is changing, a growing number of studies are being conducted under realistic conditions in recent years. In addition to the methodological remarks, research on the integration of robots into care settings has rarely gone beyond brief observations. To eliminate novelty effects, we argue that such research should be conducted over a longer time period.

However, some studies have gone above and beyond, such as those of Wu et al. and Fiorini et al. [107, 338]; both conducted studies in a living lab environment, albeit only to a limited degree with ethnographic methods. Previous research placed more emphasis on the social aspects of robots (see, e.g., [177, 266, 268]). An example is the research of Sabelli et al., who have been assessing only conversational elements, with conversation input created via a "Wizard-of-Oz" system, but also including care employees in the analysis [268]. Other researchers have looked at how users react to different robot facial expressions to explore categories such as social bonding or intimacy between a participant and a social robot [177].

Although social robots are applicable to the care environment, the development of such robots has, from my perspective, not sufficiently involved care professionals, informal caregivers and other stakeholders. However, there are a few exceptions, such as a study by Moharana et al., who involved informal caregivers in an iterative co-design processes [214]. Another line of research is aiming to design trustworthy robots for care environments [300]. In regard to this, there is an interesting study by Hornecker et al. [146]. They have worked on the triadic interaction [146], looking at the shared tasks of robots, care workers and residents. Their work is one of the very few multi-actor approaches. They show that the usage of robotic systems in work environments is dependent on multiple actors working together [146]; following this argumentation, it is not enough to develop a functioning robotic system-successful usage must also take into consideration the work processes in the environment and the preferences of the people. Accordingly, although with a different timeline, we identified a need for practice-based research to investigate the interaction and make developments based on field practices [70, 277]. We further identified a need for long-term studies that work on the inclusion of care workers. Their day-to-day practices need to be included in the development of human-robot interactions in order to gain a more holistic view.

Studies with a longitudinal approach focus either on care home residents [133] or on individual older adults living at home [27, 126], measuring participants' attitudes toward the robots and their relationships with robots. Even though researchers could show how social care workers enabled interactions between robots and residents [172] (short term) or how a moderating person is required in care homes, they did not sufficiently focus on the effects that robots had on the social care workers' daily work practices. Our study [70] revealed that some activity intervention was required and that residents enjoy interacting with the robot. But in that case, the manipulations were carried out by a researcher. There is one study that assessed the effects of a NAO robot on social care workers [139], albeit without concentrating on daily work procedures. In one of the studies of this dissertation, we researched the effects of the robot Pepper on social care workers' daily work practices [72]. Another long-term study placed a robot in a nursing home as an information system for the residents in order to tailor the robot to the users' information demands; the research was carried out mostly from a robot-centered perspective [133]. A recent study also took into account several robot systems (humanoid and pet-like robots), and worked with care workers on how such robots can fit into their daily work practices [345] Expert users have talked more about the possible long-term utility of social robots, emphasizing the significance of human affection, robot functioning, and the emergence of social communities centered upon robots [92].

Overall, the number of user-centered studies is growing. We have also contributed to this line of research and have communicated our findings to the community. However, I argue that there has not been enough work done in real-world settings with a long-term perspective. In this dissertation, I present a series of exploratory studies based on practice-based research conducted over a long term.

2.2 Context

2.2.1 Aging Population

Life expectancy in industrialized countries has been steadily increasing for more than a century [233]. Two distinct influencing elements contribute to the rising average life expectancy. A decline in newborn mortality rates contributed to the increase in average longevity in the first half of the 20th century. This began to change in the middle of the 20th century as an increase in the average age of the population fed the increase [233]. Although it is difficult to predict whether the current low birth rate will persist, it is generally accepted that in the coming decades, the average age in developed countries will be substantially older than it is now [310], under the condition that there are no unexpected events, such as wars, pandemics or natural disasters.

The so-called "population pyramid" of Germany figure 2.2, as well as continuously increasing life expectancy, show a clear trend toward an asymmetrical age distribution of German society [137], a shape characterized by increased average age as well as a low birth rate [310]. Figure 2.3 is further exhibiting this situation and shows that the number of older people is rising while the number of young people is proportionally falling.

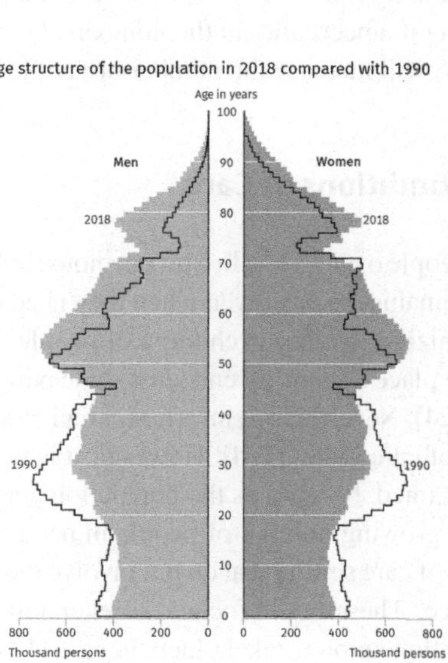

Fig. 2.2 Age structure of the population in Germany: Comparison from 1990 to 2018 [234]

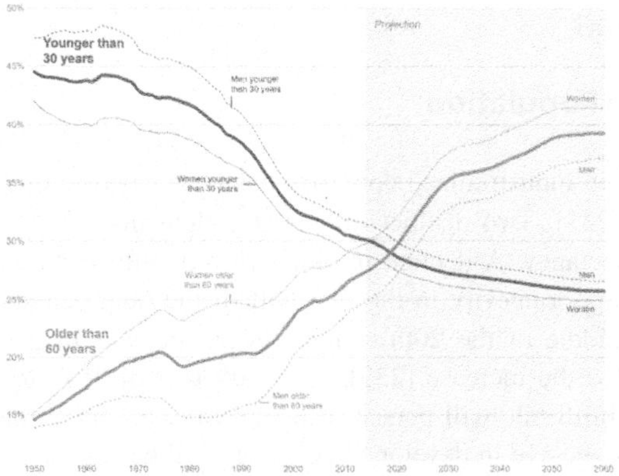

Fig. 2.3 Number of people over 60 exceeds the number of people under 30. (Source: [263])

It is unclear how this trend would affect German society's way of life. There is little doubt that these changes will have significant effects on the political and economic systems [223]. Projections by the Federal Statistical Office have shown that in the long term, Germany's population is expected to decline in the future, despite immigration [234]. Although projections are made, immigration is a contentious political issue, resulting in uncertainty in the numbers. One thing seems clear: this demographic transition will increase the number of people in need of care.

2.2.2 Societal Conditions of Care

In Germany, 90% of people over 80 live in a private household [55]. For the majority of older people, this remains the case even when they need care. It can be observed that in a mobile, urbanizing world, the children of people in need of care often no longer live in the same place as their parents, instead leaving long-term for training, studies or jobs [102, 124]. Nevertheless, informal caregivers (i.e., family caregivers) still provide the bulk of care today [124]. However, it is unclear whether this state of affairs can be maintained, because as the population ages, fewer helpers will be available to care for a growing number of people in need of care [314]. It is more likely that other forms of care setting that do not involve the assistance of the family will gain in importance. These could include weekend and night care as well as more in-depth cooperation among stakeholders in care [124].

The fairly recent development of families increasingly living apart, as well as the demographic change, is putting increased pressure on the care system and creating bottlenecks. It is not certain that the quality of care we know today can be maintained in the future, as it will have to be provided by relatively fewer employees [118]. In addition, the work load for caregivers is already growing, as people in need of care come to care homes (or nursing homes) at a later stage and therefore need more intense care, while at the same time the profession faces overtime hours and a high number of vacancies [19, 129]. Overall it is apparent that there is a dire need for assistance in care settings. This work attempts to investigate whether social robots may serve as part of that, but it should be noted that numerous components will be required to stabilize the system.

3 Methodology

In the following sections I will describe the methods deployed in producing the results of the studies in the thesis. I will further elaborates on the challenges with COVID-19 and the use of citizen science. The chapter will discuss the methods used and the opportunities and challenges they involved. The reader will obtain an overview of how I worked over the span of this dissertation. For more context, I recommend looking in the studies themselves, as it may be helpful to see the specific context in which the methods have been used. The methods, in turn, were influenced by the focus of the study, and the technology or group of people in question. The limitations of the described methods will be enumerated in the limitations Chap. 11.

3.1 Understanding Practice

In this section I describe how I learned through my own practice and how I made sense of unfamiliar contexts. The field of care was not native to me and I used different strategies to gain a deeper understanding of it.

3.1.1 Participatory Design

This dissertation comes from a Socio-Informatics [342] tradition and has at its core the elements of participatory design [325]. The concept can be traced back to political movements in the West about various political, social or civil right topics; people wanted to have a say in decision-making and be part of it [258]. It was later introduced to the IT sector, in Scandinavia it was used when computers were introduced to the work environments. This move was connected to the trade unions and was

F. Carros, *Design, Development and Sensemaking of Human-Robot Interaction in Care Settings*, https://doi.org/10.1007/978-3-658-45233-9_3

called the "workplace democracy movement," giving people a voice in the introduction of technology and its usage in their work practices [38, 230, 258, 269]. In its early use, the participatory design framework saw two main groups that it addressed. One is the "users," the people who are going to use the newly introduced artifact [258]. The other group are the "designers," who create the artifact and shape it to the needs of the users [258]. Later it became clear that there are more than one group of users; in many organizations there are several persons directly or indirectly affected by new developments, and hierarchies within a company also have to be taken into account [259]. The same is true for care homes. Although we mainly develop the social robots for residents and care workers, there are many other who are influenced by it, such as insurance companies or care school teachers. The framework of participatory design is also linked to ethnographic approaches, but there are some differences. Whereas in ethnography it is recommended to gain a holistic point of view, the participatory design believes in expert knowledge and respects that different knowledge is possessed by different persons [40]. Further, there is a difference in the participation of the researchers: whereas in ethnography, the researcher often reflects their participation in activities in participatory design, it is seen as normal to part-take in activities with the participants and to discover together [165]. But ethnographic research can inform designers on how IT developments could be made, based on their research. This becomes especially handy when it is not possible for the designers to go into the field; they can then rely on working with ethnographers to receive the needed information to make an informed design decision [149]. An example of this kind of scenario is the work with social robots during the COVID-19 pandemic. In its early weeks it was impossible to go into the field, so we were dependent on information from others as how the situation was and what development possibly could help. Although there were no ethnographers in the care homes, we used ethnographic methods such as diaries, phone and video calls, interviews, and photographs. This provided us with insights into the situation and assisted us to identify which developments might be beneficial.

There is also a political dimension to participatory design (PD). As explained earlier, PD has its roots in the democratization of decisions [258] and was then later applied in the field of information technology [170]. Under regular conditions within an IT participatory design project, it is not possible to include every person who will be affected by the development. Sometimes hundreds of people are influenced directly or indirectly, within different organizations, levels of hierarchy or different roles (e.g., families and care workers). Therefore, a group of participants has to be chosen (or they have to be willing to participate); the constitution of the group members is not representative and will have a bias. I and my collaborators tried to engage people working and living in the setting of care homes. Some of them were residents of the care home or receivers of care services, while others worked directly in the field and interacted with the residents and care recipients. We also worked with secondary stakeholders: people who do not directly work in the field but whose work is closely connected to the field and are influenced by the developments in it. The decision to use PD is linked to the sensitive nature of the field of care and of the topic of exploring social robots in care. It also reflects the tradition of the environment I have worked in (see, e.g., [74, 75, 216, 236, 274, 279, 330, 343]). We wanted to understand how a social robot can fit into this specific field and aimed to closely observe the effects it had. At the same time, we recognized that care is a highly complex field with many constraints. Ethical questions also played a role. We did not see fit to evaluate the ethical implications of using robots in care settings in advance but instead preferred to evolve an understanding of the ethical implications in a participatory manner, letting the people directly and indirectly affected decide what should be developed. We have a strong belief that people in situ understand much better how robots might be included and what tasks might be worth further developing, but also—more importantly—what robotic tasks should not be further developed. An example of this is using robots to bring non-mobile residents from their rooms to the dining area or to activity rooms. Although it might seem like a task that would help the people in care to save time and to make people wait less, we were told that this task is a social moment, one of the few where care workers and residents have time to talk to each other. Automating it might save some time, but it is an important moment for both parties that holds much more significance than just bringing a person from A to B.

For all these reasons we decided to use PD to work closely with the people in the field—not only to let them be validators of the developments, but also to co-create with them, and to explore together with them the possibilities and limitations of the robots and their possible inclusion in the field.

3.1.2 Grounded Reflections

For me personally this work was an evolving process of understanding. I started early in my studies to be interested in social developments such as demographic changes. I dived into the field during my master and regularly worked in a care home for a period of around two years and taught residents weekly how to use devices such as computers and tablets, and showed them the wide possibilities of the internet. The chair (Information Systems and New Media at the University of Siegen) had developed a format in a European-funded project that was called "Internet Coffee," in which we were supposed to teach residents how to use a healthcare platform.

They turned out to be much more interested in discovering the internet, so we spent more time than we expected teaching them how to communicate via ICT or showing them games and other applications. At the end of my master, it became clear to me that I would continue to work in this field. There were two weeks left, before my contract started at the university, which I spent at an internship in the care home we used to work with. My (care home) co-workers knew that my intentions were to better understand the field to later develop insights into the use of a robot together with them. For two weeks I had the chance to work in the field. They opened all doors to me and gave me the opportunity to ask many questions. I got insights into the routines of the care workers and followed the work not only of caregivers but also of the social care workers. It was incredibly worthwhile for my understanding of practice, as I could closely observe how caregivers and social care workers work and live, without having a special project in mind. I had shifts in the morning, during the daytime, and also at night. The second week was special; many of the residents were infected with Norovirus.

During the internship I was constantly writing in a little booklet containing my field notes, writing down my observations and experiences—but also ideas I developed and validated while working there, along with notes of the discussions about different ideas I had with the people working with me. This booklet was an important source of knowledge for the coming years and heavily influenced the developments we attempted.

In the coming years, we continued our work with the care home in a PraxLab logic [217, 235]. In 2018 and 2019 I continued to go to the "Internet Coffee," now with the intention of working with robots—but sometimes we were also just playing games on a computer. It was not possible for me to go to this "Internet Coffee" every week, but I tried to attend as often as possible. On top of this activity came the studies (e.g., [66, 70] that also took place in the care home. Later studies took place in another care home.

In these studies we worked closely with one care home, for a certain amount of time, on one specific topic, using the Design Case Study method [294, 341, 343]. Within this framework we were following the three steps of the design study by (1) doing a pre-study, (2) having a design phase, and (3) observing the appropriation of the users to the development and re-designing. In the first phase we were participating in a shared activity of the care home and the university where the residents were taught how to use tablets; furthermore, we conducted interviews with the residents and the care workers and observed their work. I also shadowed the care workers to better understand their duties and the activities they were engaging in with the residents. In the second phase we did the design of the applications. This involved producing prototypes and showing them to the residents and care workers. In the third phase we looked at the appropriation and did several re-designs until we were satisfied with the application. This also meant that we showed the participants how to use the applications and the robots. We performed long-term studies and observed how the participants were using the development, what they preferred and where challenges arose. We further tried to understand what influence the robot and its applications had on their daily practices and how their practices changed with the new development.

3.1.3 Work in COVID-19

This dissertation relies heavily on assumptions about practice and the value of working closely with residents, care workers and other stakeholders for an extensive period of time. As explained above, it was normal for me to go once or twice per week to the care home—but COVID-19 changed this. Not only was it an existential shock to all involved, but it also meant that the work I used to do had to be adapted to the new conditions. Working in care homes became even more challenging.

In the first months, the government and the care homes thought that it would be good to isolate care homes. Even the closest family members—people who were used to visiting the residents regularly—could not enter the care home. For some time, we could observe how family members (and other visitors) and residents were forced to communicate via phone or video. Sometimes, other obscure ways of communicating were used, such as talking on the phone while looking through windows. But for some residents, such as those who were bedridden or who had only limited means of communicating, even this was not possible. These residents were not always able to adapt to new ways and as a result some were cut off from communication with their loved ones.

These restrictions also affected our work and the way we could access the field: in the first months of 2020 it was not possible at all. In summer it slowly became possible again, but in autumn and winter it tightened again. We still tried to work with the care homes and wanted to document the ICT use (and succeeded in doing so, see [276]). To be able to do this, we used diaries and let residents fill in diaries daily and had video calls with the residents and care workers. In autumn 2020 we then started a field study in another care home. There it was possible to enter, albeit under strict restrictions: masks, sanitary cloaks and limited time per visit. During this field study [72] we could observe that the pandemic had not only changed the level of interpersonal contact, but also that it may have had effects on the usage of the robot. The robot was used to reduce the spread of the virus, by singing (instead of the care workers doing it), thereby decreasing the amount of aerosol. Also we saw that it was less used in group settings, in comparison to an earlier study [70].

3.1.4 Wizard-of-Oz

One of the approaches we used in the studies that helped immensely to understand the interaction between the robot and the residents was the Wizard-of-Oz method [87, 95], something I later also observed in other robot interaction studies [73]. The initial intention was to circumvent technical constraints by giving the user the impression that they were dealing with an autonomous robotic system when, in reality, a human being was in charge of the system and of what the robot said or did. This person was one of our team members, sitting in the background on the computer, observing the situation and mostly typing answers to questions that were asked of the robot or starting programs on the robot. We did not intend to use this method as anything other than as a means to bridge technical problems—but by controlling the robot, we gained a deeper understanding of how the robot was being addressed by its interaction partners. This method gave us access to the robot's "point of view," enabling us to better understand its constraints and potential. This had the bonus of making it easy to analyze the effects of alterations made to the robot. The "wizard" in charge of the robot could immediately notice changes in how users interacted with it. Additionally, it enabled us to better understand its social role in the interaction and helped us to gain a feeling of what is appropriate for the robot and what is not.

3.2 Data Collection and Analysis

To understand the interactions of stakeholders with the robot and to be able to shape the technology in order to create a better or more meaningful interaction, we relied on PD (see more details here: 3.1.1. Within this framework, we used several qualitative methods to create and make sense of data.

The most used instruments to collect data were semi-structured interviews [89, 225]. We created interview guidelines and would conduct interviews at different stages of a study and compare the data in the analysis later. Interviews varied in length between 5 minutes and 90 minutes. Typically, with care home residents they were quite short, and with secondary stakeholders rather long. This was mostly because the residents gave us short answers and often did not have the energy to have long talks, whereas the secondary stakeholders were often giving longer answers that often prompted follow-up questions. Normally interviews were face-to-face, but during COVID-19 we also held interviews over video calls.

Observations were the second source of data. Observations are at the core of ethnographic research and usually entail making sense of what people are doing in practice and their reasons for doing it that way [9, 28, 115]. In our case, it was the observation of the interactions and integration of different actors with the robot in care settings. Observations were open, and the participants knew that we were observing them. Each time, we developed an observation guideline that defined certain topics, such as interaction with the robot, problems with the robot, emotional response, effects on the group but also with the "miscellaneous" category that allowed us to write down unexpected observations. The observation protocols were filled either during the work or, if we were too busy, shortly afterward.

We also used focus workshops [6, 303] involving several people who are experts in their domain. The idea behind these workshops, and what makes them fundamentally different from semi-structured interviews, is that the participants are influencing one another. Ideas are not one-dimensional, coming from one person; instead they are criticized or enhanced by the other participants or new facets of the ideas are unfolded. I used this method mainly to elicit general opinions or identify which topics were of most interest to participants.

Videos were another element in creating data. Videos enabled us to record the interaction and were helpful when revisiting the data with people who were not present. This method of recording data was used only if we used a room regularly. Then it was possible to position the camera in a specific place. Apart from sharing them with others, the videos were also helpful to make sense of the interaction from a more distanced perspective. They made it possible to notice small details that were otherwise not seen. For example, we have seen in the videos that participants were sometimes telling very personal things to the robot—things we could also have been hearing, as we were in the same room, but did not notice at the time.

The diary method [10, 219] was particularly handy during COVID-19, when our access to the residents and the care workers was heavily limited. We used diary studies as pre-studies to discover the expectations about the social robot and how it could help in the special situation of COVID-19. But it also helped to understand the context of the residents and to gain knowledge about their feelings toward technology in general. The diary method gave us interesting insights discussed in the study that was written with Schwaninger et al. [276].

We also worked with the general public, as described in the subsection 3.3 in workshop formats, and published our approach here: [68]. They helped us to gain an outsider perspective and had an influence on the ethical and social evaluation of our developments.

To make sense of the qualitative data, we used the Thematic Analysis approach, first proposed by Braun and Clark [48, 49]. We used three methods to create themes for the thematic analysis. For one, we deductively created themes based on the interview guidelines. Second, we went through the qualitative data, mostly observation protocols and interviews, and created themes based on the reoccurring topics or based on topics that seemed important. Third, the themes were discussed with other researchers working in the group before and after the coding, and based on these discussions about the themes, other themes were created. Qualitative methods can involve a range of relations between observer and participants, from a broadly neutral "fly-on-the-wall" approach to a much more engaged "participant-as-observer" approach (see [23, 181, 333]), which is the position we adopted.

When we were in the care home, we also participated in other activities with the people in the care home, like eating together and discussing the latest news, but also talking about very personal topics such as family, love, or sickness (I will never forget how one resident asked me for advice on a dating situation he had). I also sometimes helped the residents: there were times I was helping people eat or get dressed, or when I brought them to other places. These situations were truly helpful to better understand the people and the practice conditions of the care home. It did not feel appropriate to take notes on these conversations. These experiences helped me to gain knowledge and understanding, but I have not converted them into data. Still, I believe they have helped in making sense of the existing data. From another perspective, it also felt right to do so. As researchers we often go to places and do "experiments" with participants; we then ask them what they think of it and do certain assessments. This usually takes a lot of time and effort from the participant and is mainly to the benefit of the research, but not to the immediate benefit of the participant. Being in the field and doing small tasks unrelated to the research has, to some extent, helped me morally because I felt like I was giving rather than just taking.

3.3 Citizen Participation

Because using social robots in care environments is a sensitive topic, we have been looking for ways to involve citizens in the research process in order to gain additional perspectives on the research topic and to better understand the public point of view in this regard. Our approach to integrate citizen was not like other citizen science projects, which often use citizens as co-researcher. Instead, we wanted to give individuals a better understanding of the technology options that are already available. From there, we could discuss and come up with ideas for how the robots that were shown might be employed in care settings in the future. This was crucial for us because robots might gain importance in this field of application and we believed that there should be a debate as to whether such technology should be used and if, how it should be used in future care settings. Thus, it became possible for us to exchange ideas with citizens who are not strongly connected to the field but who had a genuine interest. The aim of this participation was to gain a mutual understanding.

At the beginning of a citizen participation event, we gave a presentation on the current state of our research. This included a talk from us on the technologies applied and the practical use of these prototypes. Based on this, we had a discussion with the participants and worked on the initial questions. Then the design session began, in which we let small groups of participants work on one specific concepts for care robotics. The task was to imagine a task for the robotic system presented and to write it on a poster, using text and visual flow. Finally, these ideas were discussed in a large gathering with all of the participants. The outcomes of the design workshops and the discussion enabled us to learn new concepts and comprehend various viewpoints and arguments around this subject. Questions of data privacy and ethics were explored, along with a wide range of other very diverse and practical real-life challenges (e.g., being in contact with family via the robot or cooking together with the robot).

For us, it was crucial that these interactions with the citizen were built on a foundation of reciprocity. This implies that all parties should go away with something, and that the citizen should be given the chance to express their curiosity rather than just being subjected to one-sided questioning. Through the exchange, the citizen learned about our research, and we researchers learned about their perspective on the use of this technology. We were also able to discover a more varied and societal picture of social robots in care thanks to the outcomes of the public engagement formats, and we were able to connect our research and development with it. One important conclusion was that robotics should only be assistive to the work in care homes—they should not replace it. Robots should operate in hybrid teams with caregivers or social workers to facilitate their job and thus improve the quality of care, rather than performing automated tasks. This outcome was also confirmed in later studies (e.g., [70, 72]).

Part II
Publications

The second part of this dissertation lists six different publications that make up the core of this work. These publications are organized into six chapters, from Chap. 4 to Chap. 9. The publications are thematically connected by focusing on practice-based research with social robots and their usage in care environments.

Chapter 4 focuses on the interaction of care home residents with a social robot. Taking a long-term perspective, we examined the usage of the robot, its impact on residents, and the necessary personnel. In Chap. 5, we analyzed the care workers' perception of the robot and its integration into their daily work. After several years of development, our goal was to investigate how the robot could be sustainably utilized as a tool in the daily practices of care workers. Chapter 6 discusses a study conducted in the early phase of COVID-19 within two care homes. It explores the use of technology by care home residents and workers, their collaboration, and potential applications of robots. Chapter 7 examines the inclusion of citizens in research on social robots in care environments. Care homes are a sensitive topic that requires societal debate. Technology developments for the care sector should be designed with citizens in mind. Chapter 8 discusses the ethics of social robots, including autonomy, emotionality, sociality, and their impact on social competences. The final chapter 9 of this section provides lessons learned from working with various social robot research projects.

4 Exploring Human-Robot Interaction with the Elderly: Results from a Ten-Week Case Study in a Care Home

This chapter was published in ACM Proceedings of the 2020 CHI Conference on Human Factors in Computing Systems as *Carros, F., Meurer, J., Löffler, D., Unbehaun, D., Matthies, S., Koch, I., Wieching, R., Randall, D., Hassenzahl, M., Wulf, V. (2020, April). Exploring Human-Robot Interaction with the Elderly: Results from a Ten-Week Case Study in a Care Home. In Proceedings of the 2020 CHI Conference on Human Factors in Computing Systems (pp. 1–12).* It corresponds to the P1 in the List of Publications.

Abstract. *Ageing societies and the associated pressure on the care systems are major drivers for new developments in socially assistive robotics. To understand better the real-world potential of robot-based assistance, we undertook a 10-week case study in a care home involving groups of residents, caregivers and managers as stakeholders. We identified both, enablers and barriers to the potential implementation of robot systems. The study employed the robot platform Pepper, which was deployed with a view to understanding better multi-domain interventions with a robot supporting physical activation, cognitive training and social facilitation. We employed the robot in a group setting in a care facility over the course of 10 weeks and 20 sessions, observing how stakeholders, including residents and caregivers, appropriated, adapted to, and perceived the robot. We also conducted interviews with 11 residents and caregivers. Our results indicate that the residents were positively engaged in the training sessions that were moderated by the robot. The study*

Supplementary Information The online version contains supplementary material available at https://doi.org/10.1007/978-3-658-45233-9_4.

F. Carros, *Design, Development and Sensemaking of Human-Robot Interaction in Care Settings*, https://doi.org/10.1007/978-3-658-45233-9_4

revealed that such humanoid robots can work in a care home but that there is a moderating person needed, that is in control of the robot.

4.1 Introduction

As a consequence of to an ongoing demographic change and various associated social transformations there is an increased pressure on the care systems, which is forecasted to grow significantly [167]. Industrialized countries are confronted with a decreasing number of professional caregivers and an increasing number of people in need of care. Providing high quality care for a rising number of elderly people will become a challenge [59]. The shortage of skilled workers not only has a negative impact on the quality of professional care and thus on the physical and emotional wellbeing of caregivers, it also affects the wellbeing as well as physical and cognitive integrity of those who are in need of care. In the view of the complexity of the problem, governments, academia and industries are analyzing various strategies, including robot-based assistance for the elderly, to support their health and safety, and foster social participation (for reviews see [2, 34, 51, 52, 162, 252]).

Despite the wealth of work on social robots in aged care, few systems have yet already made their way into the real world. This may be in part because of the complexity of the care ecosystem, existing prejudices and concerns around digitalization and automation, as well as the huge variety of social practices and institutionalized routines of caregivers and residents into which any technical system must be carefully embedded. In short, there is a lack of empirical data and need for more practice-based studies that examine enablers and barriers of robotic systems in real-world care settings and over a longer period of time, taking into account perspectives from all stakeholders involved.

In this paper, we present a robotic-based application utilizing the robot Pepper that was especially designed to support older adults and their caregivers in care homes to increase physical and cognitive activity and initiate social interaction. We introduced the system in group settings with residents over a period of 10 weeks and conducted interviews and observations with residents, caregivers and a facility manager. In this way we were able to elicit these stakeholder's attitudes, social and organizational practices, expectations towards the robot, individual and group-based performances as well as social communication among the participants. We distill our findings from the long-term field deployment into lessons learned for future use of robots in aged care.

4.2 Related Work

4.2.1 Studies of Social Robots in Residential Care

The main goal of socially assistive robots in elderly care is to provide social companionship [52] rather than physical assistance such as feeding (e.g. [153]) or transport (e.g. [178]). As such, the vast bulk of research undertaken in this area has focused on the individual relationship between the robot and those with care needs. Typical tasks of socially assistive robots include cognitive support such as providing reminders of medication intake (e.g. [250, 309]), motivational work to engage elderly in physical exercises (e.g. [101]), cognitive exercises (e.g. [111]) or communication support in the form of the telepresence of caregivers, family or medical staff (e.g. [44]). The most widely studied social robots in aged care are the baby seal robot, Paro, developed by AIST, SONY's robot dog, Aibo, originally developed for entertainment purposes. Other, more complex robots follow a human model, e.g. the robot nurse, Pearl [249], Pepper, or the home assistant Careo-bot [127, 171].

Research is steadily accumulating mostly qualitative evidence but also quantitative evidence for the effectiveness of these interventions (e.g. [52]). For example, in a systematic review, Kachouie et al. revealed that socially assistive robots could potentially enhance the well-being of the elderly and decrease the caregiver's workload, as assessed from psychophysiological data like urinary tests for stress hormones and self-reported measures such as the burnout scale for nursing staff [162]. Animal-therapy robots such as PARO or AIBO, have been shown to reduce loneliness and mental stress, and increase verbal and physical activity (e.g., [163, 172, 265]). Challenges in existing evaluation studies include short study durations that can be heavily influenced by potential novelty effects, as well limited sample sizes (most studies reports sample sizes less than 10). Therefore, more robust, i.e., replicable, large scale and long-term studies in real world settings are needed to identify current enablers and barriers for robot-based assistance in practice [34].

Another factor that impacts social robots' acceptance and effectiveness in real world settings are the perspectives of various stakeholders involved. Pino et al. found that people with mild cognitive impairment and caregivers of people with dementia were much more welcoming towards care robots compared to healthy older adults [247]. This shows that current needs are a main driver for technology acceptance in elderly care, both on the side of patients as well as caregivers [51]. While most work introduces social robots in individual encounters with elderly persons, some research has shown that a group setting might be a promising alternative. For example, Kidd et al. found that the effect of a social robot increased in the presence of personnel that were willing to participate in the interaction [172]. Thinking about further adaption of social robots for everyday use in care facilities, factors such as user safety, ethical considerations such as privacy and cost effectives have to be considered also from the perspective of caregivers and care homes [86].

4.2.2 Perspectives of Caregivers and Care Homes

The profession of caregiver is confronted with a growing number of challenges in times of demographic change. Not only does the number of medical and psychological tasks increase with the ever rising age of care home residents, the profession also faces low payment and increasing stress despite the high number of vacancies [18]. This results in a situation that many care homes do not have enough staff, and this situation forces caregivers again to do overtime work [129]. Moreover, in the public opinion, the job of caregivers in care homes is often seen as relatively unskilled labor [145] that could be eventually (partially) automatized. As such, care homes and caregivers have a rather ambiguous attitude towards automatization efforts as researched in the field of social robotics, which have to be empathically addressed in field studies.

Nevertheless, both stakeholders are also open towards technological innovations that potentially help them fulfill their duties in a way that reflects the possibility of job satisfaction and that corresponds to their own ethical attitudes [155]. Previous research has shown that the development of technology in this ecosystem is better done with insights into the lives and daily practices of residents and caregivers in a specific care home, something that typically requires a qualitative approach of data collection [216, 315, 316, 318], as the adaption of technology is highly dependent on the context of use as has been pointed out many times in Human-Computer Interaction (HCI) research.

When developing for older adults and the stakeholders engaged with them, the use of 'Living Labs' has become more common in the research field of HCI. Living Labs are either build environments that are close to the real life environments of the participants. Studies there are mostly short-term that go from several hours to a few days [3]. Another approach is to work in situ, i.e., in the places participants live and conduct the research and testing as integrated part of real-life social contexts [218]. Particularly in the environment of care homes, which is a highly sensitive field, the opportunity to engage with various stakeholders over an extended period is invaluable. A Living Lab can provide useful information about the social practices of the participants and creates a situation that is close to a real life environment [86]. It allows testing technical artifacts like robots in an intimate atmosphere and can lead to a more natural usage of the developed technologies [86]. This helps researchers and developers to understand users and their requirements in a contextual way.

In this paper, we report on a ten weeks field study in a residential care home using the humanoid robot Pepper. The robot provided physical and cognitive exercises, as well as social activation to the residents. The goal of the study is to identify potential enablers and barriers of using a robot-based assistance over the long term, obtaining the views of different users and stakeholders, including the older adult residents, the professional caregivers and the management of the care home. Therefore, our approach works with several stakeholder groups and takes a more holistic point of view on field deployments of social robots in aged care.

4.3 Method

Our study was conducted in an institutional care facility with older adults in Germany. The care home had already experience in working with researchers from other ICT based projects. It is a facility with 119 residents. The aim of the study was to investigate the use of the socially assistive robot Pepper in real-world daily life situations of older adults and their professional caregivers in an institutional care setting.

4.3.1 Study Design and Setup

In the run-up of this study, we designed and developed training applications for Pepper that had been identified based on prior long-term participative interaction processes between the researchers, the older adults and the professional caregivers. The detailed overview of the technical system is described in the following section. We note, however, that the identification of the particular technical requirements of the system is not, even so, the main theme of this paper.

To explore the use practices of the participants in an institutional care setting, our study design included three steps. We started with initial interviews with all older adults and caregivers in the care home to explore the expectations, attitudes, feelings and current exercise patterns of the older adults and the caregivers. Secondly, a trial was conducted where the participants interacted with the robot system over a course of 10 weeks, with two workshop sessions every week i.e. 20 sessions in total to gather insights into use practices in a real-life care setting. Each session lasted between 45 and 60 minutes and was conducted by a research team consisting of two researchers. Each session was conducted and moderated by one researcher who also assisted the older adults when needed. To improve the user experience, a "Wizard of Oz" approach [87, 203] complemented the setup, whereby a second researcher controlled the robot from a nearby but unobtrusive position during all courses. This enables a smooth transition between applications where necessary. The performance and usability of the system was continuously assessed and re-designed based on observations of its use and feedback given by the participants. During all study phases, stakeholders were encouraged to act as co-designers, giving advice and making suggestions as to how to progress. In a third and final phase, we re-interviewed participants to ensure consistency in reporting the study results.

4.3.2 Data Collection and Analysis

Both the initial and final interviews were semi-structured, but questions were individually tailored where interesting aspects could be examined in more detail. Each interview took about one hour. The guideline of the initial interviews included questions about daily life and work in the care home, the existing approach of exercising and the participants' attitudes, feelings and expectations towards the robot system. The post-interviews focused on the service use and related experiences. The participants were asked to compare their expectations before the study took place with their actual user experiences. We also asked them to describe positive and negative experiences with reference to specific situations, as well as their wishes and claims for possible redesign for long-term usage. Interviews both before and after the field study were fully transcribed and analyzed.

With respect to the workshop sessions, the researcher who moderated the robot course sat beside the older adult participants at a table. Another researcher sat in the back of the participant group, controlling the robot with his computer and writing the field notes. In addition, every session was recorded by a camera in the back of the room. The videos helped us later to analyze the interaction and to reconfirm the field notes. The protocols and transcripts were analyzed by using a thematic analysis approach [48]. Based on the transcripts, researchers performed an inductive analysis of the data, coded it, and generated a series of main themes collaboratively. Coding discrepancies were discussed and eliminated by adding, editing or deleting themes, according to the outcome of the discussion.

4.3.3 Participants and Setting

The participants included in the study were six older adults who were residents of the care home (they will be called 'older adults' or 'residents' in the following study), four caregivers and the former manager of the care home (see Table 4.1). For the study, we received the ethical approval by the ethical committee of the University. All participants received a detailed explanation of the study aims and protocol and signed informed consent documents about their participation.

Table 4.1 List of participants (*dropped out of the study early)

ID	Name	Sex	Age	Role
1	Mrs. M	f	94	Care home resident
2	Mrs. T	f	84	Care home resident
3	Mrs. H	f	91	Care home resident
4	Mrs. E	f	88	Care home resident
5	Mrs. S	f	80	Care home resident
6	Mr. V	m	91	Care home resident*
7	Mr. H	m	62	Caregiver—Social Service
8	Mr. E	m	65	Former Manager
9	Mrs. B	f	48	Caregiver—Social Service
10	Mrs. A	f	35	Caregiver—Social Service
11	Mrs. W	f	62	Caregiver—Social Service

All six older adults have been living in the care home since for some time. For the study, we chose persons that were not diagnosed with any type of dementia or MCI. Nevertheless, they had different impairments. Most participants were in need of visual and mobility assistance and one resident [Mr. V] was in need of a hearing aid. It turned out that even with his hearing aid he was unable to understand the voice of the robot well enough. For this reason, he dropped out of the study at an early stage and the study carried on with five female older adults participants.

The four caregivers and former manager of the care home attended the workshop sessions and had conversations with the older adults. During the workshop sessions they mostly sat in the background. They were also an important source of knowledge for us in understanding more about the context of the care home and the former exercising opportunities and habits of the residents.

4.4 System Overview

4.4.1 Technical Infrastructure

Pepper is a humanoid robot that is 1.2 meters high and weighs 28 kilograms [243]. He is equipped with 2D and 3D cameras and has microphones as well as a tablet for human robot interaction. He has two arms that can separately be programmed to do movements. He is anchored on three rollers and can move around with them on firm ground. The robot can also move his head and hip, and the face looks a bit like a human child with very big blinking eyes (see figure 1).

Using microphones, the robot can interpret the spoken word of humans. However, the speech recognition algorithm is limited and often does not work correctly. This occurred particularly often in group settings, with several persons speaking at the same time or with persons not facing the robot while speaking. To overcome this technical limitation, we used a laptop and a smartphone that are connected to the robot and allow intervention on the operator's part, like in the Wizard of Oz approach [87, 203]. In case the speech recognition failed, the spoken words of the participants were manually entered in the laptop to provoke the desired reaction of the robot. Furthermore, a smartphone was used to make the robot say phrases that were not fixed parts of the program. For example, to motivate a specific person, the robot said things like *"come play with me, Mrs. E."* Although participants knew that the robot was partially controlled by the human researchers, they did not know at what point specific interventions took place. The programs on the robot were developed by our research team following a participatory design process. Together with the caregiver and the inhabitants of the care home we have thought of applications which could fit into this setting and have developed them. Through many iterations with them we arrived at the program which has been used in this study. Some parts of the program were as further adapted during the study, based on the needs and continuous feed-back of the users.

4.4.2 Robotic Intervention Course Setting and Concept

The course took place in a common area room of the care home that is usually used as a meeting room or cafeteria. Participants came for a ten-week period, twice a week at a fixed day/time. Two researchers, one working with the robot and computer, and another one facilitating the interaction with the users and explaining the interventions in case of difficulty, moderated it. The six older adults' participants sat in a semicircle in front of the robot as seen in Figure 4.1. This setting was used most of the time and only changed when a table was needed, while the caregivers took only part sporadically in the sessions.

Fig. 4.1 Group setting with the Pepper Robot

A first general concept for the implementation of the intervention was developed based on an initial requirement analysis, interviews with residents and caregivers, and field observations in the care home.

The final concept consisted of three phases, which were run through one after the other: (1) *motivational phase*, (2) *physical activation*, and (3) *cognitive activation*.

Motivational Phase

The first phase focuses on slowly getting the participants used to Pepper. To ensure this, it starts with a music related game. Music has a high value in the life of the participating older adults. By using well-known applications and songs, the threshold of the older adults participant's engagement with Pepper was reduced. We chose applications in which the participants do not necessarily have to interact directly with Pepper but have the chance to get used to the robot. Apps such as a music quiz were used, where the residents get to listen to parts of a song and then have to guess the interpreter or the name of the group. In case a good answer was missing, Pepper helped with additional hints. If the speech-recognition misunderstood the answers, we typed in the answers in the program, like a Wizard of Oz.

Physical Phase

The main phase of the intervention is the physical phase. In this phase, Pepper's humanoid appearance is used to demonstrate exercises to the residents. These exercises are based on existing exercise programs like OTAGO and FAME that have proven effective for the health of elderly people in various studies, as well as exercises learned during the initial requirement analysis [11, 96]. Pepper guides and explains the exercises and act as an exercise trainer. Due to technical limitations, Pepper is not able to ensure the control and correct execution of the exercises and movements of the older adults. Pepper emphasizes regularly that the exercises should only be done as far as it feels good and feasible for the participants. Still we had to control the health condition of the participants during that phase, in case someone would get too exhausted, but the program was designed to use submaximal exercises. The phase itself is divided into four smaller parts: warm-up, flexibility, strength and cool-down, each lasting about 5 minutes; the parts have been started from the researcher on the laptop.

Cognitive Phase
The final phase of intervention activates the older adults cognitively with a playful approach. In order to present the participants a varied experience, different games have been used to maintain a rewarding character of this phase and to motivate them for their further participation in the project. The games are designed to promote memory, reaction or concentration. Furthermore, the games used were designed to encourage the participants to discuss in the group and reunite them as a group after the second phase. One of the games is a quiz in which the robot acts as moderator and the participants are separated into two groups to compete against each other while needing to discuss within each group. In this phase, it became necessary to move the robot from one person to the next.

4.5 Findings

Using a thematic analysis approach, several themes emerged during the process: the context of individual and group-based Human-Robot-Interaction, positive social experiences and cooperation between the participants, attitudes and fears towards the system as well as social and ethical issues. In the following, these themes will be used to structure the results section, illustrating how the robotic-based system initiated a range of impacts amongst participants and their caregivers.

4.5.1 Getting Engaged with the System

The workshops included different kinds of activities that aimed to support the physical and cognitive abilities of the participants by getting them engaged with the help of the robot. During the explorative workshop sessions, we could identify many factors how that had an influence on the participant's engagement.

Getting familiar with the system
An important aspect was that the older adults needed to get familiar with the different kinds of activities and the robot. For example, it was a recurrent pattern that the older adults were more restrained and insecure in the very first sessions, while they became more actively when they started to get used to the new setting and the robot. That could be observed e.g. in the activity 'song guessing'. While many of them were still quite reserved in the first session, they began to participate actively on their own from the second session onwards. As soon as Pepper started singing songs that they knew, they began to sing along quietly, to move back and forth slightly or to move their feet slightly to the music rhythm. The older adults have guessed many of the songs that have been played by Pepper correctly and the activity encouraged an exchange about whether songs were known among the participants or not.

A similar observation was with another music-based activity. While the older adults needed some extra support by the researchers to start singing or moving rhythmically to the songs, this started to change already in the next two sessions, when the older adults at participants get familiar with the format of the activity. From the second week onwards, all physical exercises were executed as explained by the robot and could be embodied by it without further introduction or guidance of the researchers. After a few weeks, the additional videos for explaining lower limb movements were totally removed because the residents knew the expected exercises and were familiar with the tasks. Until then, then older adults' were very actively engaged and mainly focused on the robot during the execution of the exercises but also watched what others were doing when the robot explicitly praised individual participants.

Hence, the two examples also showed the importance of getting a better understanding of the needs of this particular group of participants. Thus, we noticed that the engagement was much higher when the older adults knew the song. Unknown songs did not lead to any movements except when the robot asked the participants explicitly to do specific movements along with the song and demonstrated this with his own movements. As we noticed at this point, we implemented a function whereby Pepper asked for their favorite song to optimize the selection for the next session. The robot 'promised' to practice new songs for the next session to make the sessions as varied as possible and to suit the group's taste.

Overall, the older adults gave us a positive feedback with regard to the study and told us that they felt positive about the interaction. Mrs. M. had the following thoughts about it: *"I liked it all. But in the beginning, you have to, you have to dig in first, of course."* Mrs. H. has made similar experiences: *"It's always been interesting. [...] he was good to understand [...].* She even told her son about it: *"I told my son about it, then I always said I would go back to the robot today. And then he said yes and laughed"*. It was a pleasant course for her; she added that if not *"I wouldn't have come. Then I would have said I was sick."*

Dealing with accessibility issues
During the study, we noted that the participants had to deal with accessibility issues when interacting with the robot. The voice of the robot was a problem. One resident, as stated, left the group, since to his loss of hearing created difficulties. The ability to actively engage with Pepper varied across the participants. The video analysis revealed that some participants were more able than others to follow the instructions by Pepper and to imitate the movements he was demonstrating.

In the videos, it became evident that there were on occasion certain obstacles to common interactional quality. In the course of the different activities, and in some cases, Pepper's instructions had to be repeated several times, sometimes if the robot spoke too fast, or when the applications had to be described if they were too complex. However, it also appeared the residents become more used to Pepper's intonation after a couple of sessions and were thus better able to anticipate instructions.

4.5.2 Attitudes and Feelings towards the Robot

During our evaluation phase, we saw evidence of positive feedback between the older adults. Particularly in the first sessions the older adults showed considerable curiosity about the robot. They looked at him and touched him. They explored his hands and touched his head, sometimes as if you would pet a cat or a dog and started to explore on his reaction. The residents also talked with him, called his name and asked him simple questions like *"how are you today (video 3)"*. The participants were intent on very simple language and often spoke in a very friendly manner. The bodily and verbal reactions of the robot often made the participants smile and even laugh. The attitude towards Pepper was open-minded and rather curious. Thus, Mrs. SL. attended a session as a visitor to see Pepper once: *"I just want to see him and watch him (Pepper)"*.

However, the interactions did not always turn out that well, and sometimes even became a little awkward. Often the older adults asked complex questions like: *"It's nice that you're back again today. Are you happy too?"* In such situations, which the robot had not been programmed for, it answered e.g. *"Excuse me. I didn't understand that."* When the participants then repeated the question, the robot repeated its answer in exactly the same way. This often led to a certain resignation by the residents, because the limitations of the robot's capabilities became evident. Hence, when we asked the participants how they feel in such situations and if this causes some frustrations on their side Mrs. S responded (in video 11) that it does not bother her at all. As with pets and children. We found that the participants were very patient with the robot. When an error occurred during the intervention. Mrs. S answered in a forgivingly way: *"Well, the little guy has still to learn"*.

There were examples of humor, when Pepper said something that they deemed inappropriate, phrases that did not fit the current social situation. In those cases, the interaction often drifted away from a focus on Pepper and became a shared interaction between the participants. In addition, when enjoyment was evident, it was manifest in joint laughter.

Nevertheless, in the interviews we discovered there was also some concern about the robot. Caregiver M. raised the question *"[…] what happens if he has a malfunction, is he then just driving towards people? […] What could really happen if nobody is around?"* In addition, some participants of the study told us that they would not be very comfortable if the robot would drive around the care home all the time. For Mrs. H. it would feel like *"[…] the world would be an unsafe place."* However, in time that feeling changed *"now it is ok, now I know him, but in the beginning, it was frightening."*

Caregiver A. told us that a participant in the study talked with her about the robot and revealed to her *"She [a participant of the study] told me that it is like a friend to her now. She really has a lot of fun with it. In the beginning she was afraid of it, but now she has a lot of fun with it."* Most overcame their fears over a short period. Mr. E. the former manager has observed a shift in the feelings towards the robot. *„I remember two years ago, when the Pepper came for the very first time, it was in our restaurant, she [a participant of the study] was sitting far away and looking at it. She said, "No, I have to go now." She was afraid in the beginning. But now she is in your group and wants to be a permanent part of it."*

In the interviews with all participants the process of appropriation came up several times. Caregiver M reports: *"[the participant] said at the beginning "I enjoyed it, but it also frightens me a little". Then I said, "What are you afraid of?" she says, "It's a machine, I know that. I am not stupid. And then the big eyes […], but now, now it works, now I know him, […] but in the beginning he scared me."*

Caregiver A had a similar experience: *"[…] I observed that the inhabitants become really familiar with it. […] they really find such an access. In the beginning [they said] "hmm", "what is that", "what's new" but now I think the residents have accepted that quite well.* It is important to be there on a regular basis for a longer period of time, said caregiver B: *"[...] if there is a regularity behind it, if you not only come every 3 months or so, but if he is always present. [...] because the old people forget then. [...] So then again and again they must be occupied and shown."* On the question whether the interaction with the robot was still difficult the participant Mrs. E. answered: *"No, now it is good"*.

In the interviews with the older adults, we asked if they see the robot as a creature. We received different answers to it: *"As a creature? No, then he would have to be able to walk. If he could walk, then yes"*. Another participant has a special focus on the fingers of the robot *"What I like about this chap, has always been these fingers. [...] I think they are beautiful."* Mrs. M. sees something in the robot *"of course, he has his own character."* However, not all the residents see it that way. Mrs. H. has a different opinion about it, which shows how he understand the robot: *"Isn't it depending on how you adjust him?"*

4.5.3 Social and Ethical Issues

During the study time, some caregivers addressed the topic of the replacement of workforce. Especially for the caregivers this has been an important discussion point. Caregiver M. has a clear opinion about it *"He cannot replace us. We are 100% sure."* This opinion results from the observation that *"[...] there always has to be someone around that pays attention."* Moreover, from the feedback they receive from inhabitants of the care home, they are missing something: *"He lacks a heart, he lacks warmth." "[...] the last thing we want is that there are only robots around and you [the caregiver] are not employed here anymore."* Caregiver M see it more as a topic of the future *"I think in [...] five years something like this will exist. [...] it will be in a common area like here, with chairs and the robot in the middle of it. The robot will be activated and then will occupy the participants for one hour or so, I'm convinced that in a closed room something like this is possible."*

For Mr. E., the former manager of the care home, there will have to be a symbiotic relationship. He sees two groups in the care home *"[...] one group that doesn't want to interact with the robot, they don't see it as something serious and then there is a group that like to do so."*

He has seen a shift in the concerns of the caregiver *"[...] in the beginning there have been fears, also from the workers [...] [they knew] robots in the economic system replace them in workplaces and that's what they thought. [...] They now have learned what he [the robot] can do. Now the caregivers are not afraid anymore and the social service is not afraid anymore, they know that there needs to be someone who controls the robot, who takes care of him. And that's where I see the problem, you need trained staff and the people that are nowadays in the social service are not comfortable with technology."*

On the question of whether the course would have been different if a real person had conducted it Mrs. H. answered, *"I don't know. Maybe the person would have spoken more."* Mrs. M. has a similar opinion *"[...] you cannot compare it to a human."* The participants of the study do not see the robot as a replacement for the human care workers. Mrs. T. has a clear opinion about it *"It is better, when people do it."* Mrs. S. confirms this aspect *"I mean a person is something different. We all know that."* She sees different kind of applications for the robot that could be useful *"He can do other things, like tell you what you are looking for in the shop or show you the way or reach a hand. Nevertheless, in the care he will never be used. If it would come to it, I would be irritated. That he should wash me, or something, no! [...] It is a weird idea. I cannot become comfortable with that idea. Nevertheless, he [the root] could wake me up in the morning, he could come here [in her room]. That I would not mind. [...] Playing with him, that would be another option."* Mr. H also sees alternative tasks for the robot, for instance, *"He could read from the newspaper, if someone has difficulties to read by himself."* But there are also limits, as Mrs. A. points out *"I cannot imagine a robot during a church mass"*.

It was also noticeable that the role of the human actors changed in the course of the sessions. Early on, the caregivers had a larger presence, but their participation decreased after the third session, visiting only sporadically to check the attendance and to see if everything was working satisfactorily. In effect, this was a function of increased trust, and the caregivers were able to use the time for other tasks, e.g. to speak with or help other residents outside the room. Hence, the older adults reported that they enjoyed the time with the researchers. Mrs. E. said in one session: *"I think it's nice that young people are doing this to us old people today. I really find that beautiful"*. Thus, it might be the case that personal contact with the older adults is still an important point in facilitating motivation and engagement in the sessions.

4.6 Discussion

In the following, we will discuss the findings of the study we conducted and provide lessons learned as well as limitations.

4.6.1 Accessibility of the Robotic System

An important factor in getting the participants of the study used to the robot was to establish a certain routine and to include persons that they knew. These results are in line with the results of Kidd [172] who found that the participation of older adults increases when including caregivers. Our study took place for a period of ten weeks and has always been at the same time on the same two days of the week. The residents knew exactly when the time for this course would be and even waited for it. One caregiver reported us that this regularity has helped the older adults to get used to it and to build up trust. To conduct this research in a living lab as described by Chi [77], it was important to build up this level of trust. Furthermore, the residents were asked frequently what they liked or disliked about a function of the robot. This participatory design approach is consistent with research in other care homes, even without the use of robots [216]. Having said that, we sometimes found that the older adults did not always express their likes and dislikes directly, tending instead to talk about, for instance, music and other interests. This was scarcely a problem since understanding participants' interests enabled us to make informed decisions about the robot intervention programs, which we adjusted continuously during the study. The robot and its behaviors became something that the older adults helped to develop and represented their ideas of how it should work. Again, this is consistent with the work of Lee [186].

Nevertheless, for all the participants, it was not always easy to predict what the robot was doing. Its actions often remained unclear to them. The robot was sometimes moving with his hands or moving around in the room in ways that participants found difficult to account for. These features are programmed and are intended to make pepper more humanlike, but when the *reasons* for a certain behavior are opaque, users are reminded of its robotic nature. This barrier is difficult to overcome. Judgements concerning how humanlike a robot should be do not depend on physical movement alone. Movements have to be interpretable. Interactional quality, in our view, depends precisely on the degree to which this is the case. Previous studies have shown that humanoid design often results in a higher degree of 'likability' [120] but arguably interpretability is equally important.

4.6.2 Attitudes Related to the Robot and Course Setting

The findings of our study show that Pepper can quickly become familiar to the residents of a care home. In the beginning, the caregiver and older adults reported that they were reluctant to use, and sometimes even afraid of, the robot. They did not know how the robot worked and sudden movements or situations in which the robot was driving towards one person frightened them. However, these feeling changed quickly when they got to know the robot and got a feeling for how the robot worked. Some residents even called him by name or petted his head. simple and friendly interactions reminded us somewhat of interaction with very young children or even pets and is consistent with previous studies which demonstrate a changing orientation to technically limited applications [163, 172, 221]. Some of the older adults told their relatives about the robot and discussed it with them. Others talked with residents that did not take part in the study about the sessions and the robot. These two examples show that the robot became an object that the participants liked to talk about and was something that they thought about. Nevertheless, the robot remains something that they did not fully understand. They saw it as a machine, but as a complex one that was somewhat opaque to them. The things the robot did were often surprising for them. One of the reasons for this might have been because they did not have the possibility to customize the applications of the robot which made inferring the reasons for its actions more difficult. As discussed by Pino it is crucial to consider the needs and expectations of the caregivers and older adults [247]. Therefore, the possibility for caregivers to individualize and change the applications is likely to be productive. The end-user development method described by Liebermann et al. [191] could enhance the interactivity and transparency of such a system, since the caregivers and care home residents know their needs best. This extension would possibly have the potential to overcome the transparency otherwise lacking and might result in a deeper understanding and extended use of such a system.

4.6.3 Social Consequences

The study has shown that the residents of the care home in which we tested the robotic system are willing to receive guidance through a robotic support system together with a human. They follow the commands from the robot and imitate the movements the robot is showing them. The robot was able to get attention from the residents for the duration of a course, but it was not able to function completely autonomously during that time. The aid of humans was needed by the robot and desired by the participants to make the course work smoothly. This lack of automation might lead to an expectation mismatch, as argued by Breazeal [50]. The robot with his humanoid appearance looks like if he could work autonomously, but in reality, needs the help of a human to work in complex social situations.

Some major constraints prevent the solitary use of such a robotic system in this setting. For one, even though the applications were designed without the intention to use the "Wizard of Oz" method [87, 203], the robot was not always capable of understanding and responding to the voices of the older adults correctly. It became necessary to type the answers of the residents into the program via a smartphone in real-time. Secondly, the robot does not know the position it has in a room. That makes it difficult to move from one person to the next. In the game sessions it was necessary to be in front of the robot. The moderator had to move the robot from one participant to the next participant so that they could read or see what was shown on the tablet. This problem might be solved in the near future, since there are research groups working on this [121]. Nevertheless, this remains one of the challenges in this study. Thirdly, the older adults were willing to engage in the interaction but sometimes situations arose, in which they have shown themselves to be reluctant in trying out applications. Trying things out was often the result of acts of persuasion by a moderator, or after watching others accomplish tasks. In many cases, they changed their mind after a first interaction and were happy to use the application at the second time and until the end of the study. It seems that the personal connection between the participants and the caregivers or researchers played a role in persuading them use the system.

The last point concerns the norms and values of the residents, who were quite clear that they did not see the replacement of human caregivers by robots as feasible, likely or desirable. There was, however, acceptance of additional services, like this course, which supplement human provision. Robot intervention, in short, was seen to be desirable, when it enhanced the level of activities in the care home.

4.6.4 Ethical Issues

Humanoid robots are often depicted in a language which humanizes. They have often been referred to as "he" or "she" and given names. Even in natural sciences, humans sometimes use biomorphic descriptions [90]. Humanoid robots specifically designed for interaction with humans are usually intended to remind us of living beings. An example from the animal world is the Paro robot, whose appearance is reminiscent of a seal. Paro was specially developed for persons with dementia. It interacts with people in a simple way, for example by reacting to physical stimuli or demanding to be touched [265, 323]. Robots of this type are sometimes referred to as emotional robots [26]. Their use is aimed at satisfying psychological needs, which include, for example, the need for society and the need for commitment [26].

Pepper is a humanoid robot, its shape being inspired by a human child's appearance and it is able to move some parts of his body in a manner which apes human movement. Some early results have shown that it can enhances pleasure for people living in a care home [315]. These results are in line with the findings of our study. Pepper was accepted as a training instructor and guided them through different physical and cognitive activities largely to their satisfaction. Interviews with the participants made it clear, however, that at all time they knew that they are interacting with a machine. The humanoid design, while having some affective consequences, did not in itself enable participants to draw inferences about behavior in the way that they would with other humans. Robot behavior, it seems, remains partially unaccountable, perhaps because of the absence of 'repair' possibilities. The users are unable to adapt the system by themselves and therefore have a reduced understanding of how the system works. Such possibilities can, of course, be introduced with more sophisticated prototypes.

Another ethical challenge that arose during our study, was the lack of sustainability. With this specific group of older adults, we conducted the study during a 10-week period. Further we did some interviews before and some after this period, but nothing substantial happened after that. The study was implemented as an activity for the inhabitants of the care home. Usually these activities are designed to be offered for an indefinite time and only terminate if there are not enough people interested or other circumstances makes it necessary. An exercise and gaming course like we did with the robot is something common in the care home. They are offered by the caregivers or by volunteers. During our study we built relations with the older adults and got to know and like each other. Further the residents got used to doing such an activity twice a week. Even though we included the caregiver in the process, it was not possible to maintain the activity after the end of the study. While this might be normal for research projects, it is unusual for care homes to participate in such studies and left them with a gap to fill. In passing, we should note that a similar situation arises in respect of relations with researchers who built trustful relationships which could not, nevertheless, be maintained indefinitely. On the other hand, the care home could also offer other courses -without robot support- in which the older adults could train and perform physical, cognitive and social activities together with other residents and the well-known care staff.

4.6.5 Lessons Learned

During this study, we learned a few lessons that we believe might also help other researchers and developers to do further work in this area. Eight lessons seem significant to us.

(1) The older adults enjoyed the interaction with the robot and engaged with the system during the 60-minute sessions. During the course, the participants accepted the robot as a training guide. They were willing to receive guidance from the robot and followed its instructions in physical training and gaming situations, although human moderation was occasionally necessary.
(2) The older adults made it clear that they do not want robots to replace caregivers. They like the interaction with the robot, but it is something additional for them. Caregivers can accept such a robot as a support tool for specific tasks of activation, but not for real, intimate, or medical care work. The work that could be done by the robot belongs to other support services.
(3) Having a group setting has had a significant impact on the engagement with the robot. Often situations arose in which individual older adults did not know what to do, because they did not understand the robot completely. In these situations, they relied on each other to make sense of robot behaviors. Furthermore, the group often laughed together, something that would probably not have happened if they were alone, implying that social bonds played an important role.
(4) Including the caregivers of the care home in the course setting and, in the development, has enhanced the trust of the participants in the study and in the robot. We believe that without that, the engagement of the participants would have been significantly lower.
(5) The robot Pepper showed several technical limitations for older adults that have resulted in a reduced accessibility for them. The problems are in regard to the speech recognition, navigation and the touchscreen. The limitations made it necessary for some control of the robot by a moderator or by wizard of oz.
(6) Participatory design and the living lab approach is an essential prerequisite for understanding the needs, attitudes and practices of the older adult participants. We gained significant insights into how such a robotic system should be designed and what purposes it might serve integrated in this specific environment.
(7) For the residents and the caregivers, the robot and its algorithms are not transparent. For them the robot is a complex technical device that is not under their control and where accounting for its actions is not always straightforward. The

end-user development method which was reported by [191] could possibly fill this gap in the future.

(8) A moderating person is specifically needed to facilitate the interaction with the robot; sometimes the functionality of the applications need to be explained and the robot needs to be moved from one place to the other or someone is needed to motivate the participants to interact with the system.

4.6.6 Limitations

The participants and the researcher have built a close relation to each other. Some of the participants were in tears at the end of the study when the researchers left the care home. Therefore, it could be that some of the positive effects cannot be solely attributed to the interaction with the robot. Some of the reactions might have been triggered by a kind of 'Hawthorne' effect [257].

From a technical point of view, we had some limitations regarding some malfunctions of the robot. The user experience might have been disturbed by this. It sometimes happened that the robot applications did not load in the timeframe that was intended or that the sensitivity of the touchscreen was unsatisfactory. These technical issues might have affected the overall experience.

Despite the careful selection of the participants of the study, a general transfer of the findings to a broader population cannot easily be made. The sample size and diversity of the participants is not significant enough for this. The findings should be seen as preliminary. The lessons learned are valid for our group of participants and the specific setting. It remains unclear if the results could be applied for other groups or settings. Further research is clearly necessary.

4.7 Conclusions

We observed that the older adult participants of the study are willing to receive guidance from a robot during a physical and cognitive activation course twice a week over a period of ten weeks. This is a role that typically has been filled by caregivers or volunteers. Nevertheless, the degree of automation of such a system is far from being complete without guidance from a human, and the participants stated that they feel more comfortable in a group setting, together with human care staff that they know personally. Therefore, humanoid robots cannot -at least at present- be something that replaces caregivers, and it remains clear that the acceptance for this is low. Nevertheless, carefully designed humanoid robots like Pepper with the designed software from us, constitute a valuable diversion and entertainment support tool for serious gaming and for prevention goals like physical or cognitive activation in elderly care. Such preventive activities are demanded every day in elderly care, and the lack of trained staff is problematic, especially in face of the demographic transition and associated consequences for future care systems. Furthermore, such a robotic system, if appropriately designed and put carefully into real-life practice, can have positive impacts on some of the care home inhabitants, and thus serve as a support tool for gaining more individual human-human time in such a restricted care setting, since this would give the caregivers more time to interact with persons in need of care.

For the future and as an outlook from this work, we would need to investigate larger numbers of participants in different settings. But first, human robot interaction designers need to enable more functionality, transparency and adaptability of such robot solutions, in order to make that feasible in a broader range of different care home settings.

4.8 Acknowledgments

We thank all the older adult volunteers and care home staff, who participated in the study. The study researchers and participants did not receive any funding and did not have any conflict of interest.

Care Workers Making Use of Robots: Results of a Three-Month Study on Human-Robot Interaction within a Care Home

5

This chapter was published in the Proceedings of the 2022 CHI Conference on Human Factors in Computing Systems as *Carros, F., Schwaninger, I., Preussner, A., Randall, D., Wieching, R., Fitzpatrick, G., Wulf, V. (2022, May). Care Workers Making Use of Robots: Results of a Three-Month Study on Human-Robot Interaction within a Care Home.* It corresponds to the P2 in the List of Publications.

Abstract. *Research on social robots in care has often focused on either the care recipients or the technology itself, neglecting the care workers who, in and through their collaborative and coordinative practices, will need to work with the robots. To better understand these interactions with a social robot (Pepper), we undertook a 3 month long-term study within a care home to gain empirical insights into the way the robot was used. We observed how care workers learned to use the device, applied it to their daily work life, and encountered obstacles. Our findings show that the care workers used the robot regularly (1:07 hours/day) mostly in one-to-one interactions with residents. While the robot had a limited effect on reducing the workload of care workers, it had other positive effects, demonstrating the potential to enhance the quality of care.*

Supplementary Information The online version contains supplementary material available at https://doi.org/10.1007/978-3-658-45233-9_5.

F. Carros, *Design, Development and Sensemaking of Human-Robot Interaction in Care Settings*, https://doi.org/10.1007/978-3-658-45233-9_5

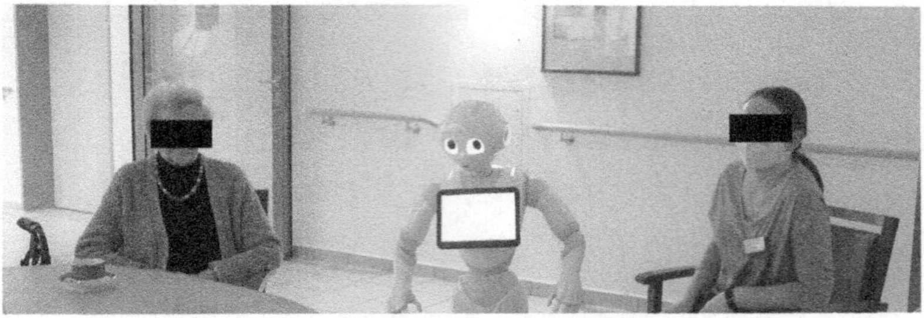

Fig. 5.1 Care worker and resident sitting together for coffee with robot

5.1 Introduction

Research into the potential of social robots in care settings is by now commonplace. It has been argued that the importance of robots in the future can only increase because of the increasing pressure on resources in care settings. At the same time, and despite this, there are very few examples of such robots becoming part of the ordinary, everyday, work practices in care homes and related institutions. The pressure on care settings is marked by a situation with a constantly increasing number of people in need of care and a non-linear growth of people willing to work in this sector, resulting in a shortage of workers in this sector [59, 167]. They are the main subject of our study.

The three month study we report on below addresses the question: "How can social care workers make use of social robots in their daily work", by placing a social robot in a care home for an extended period of time. The question specifically arose during the COVID-19 pandemic as it was thought that robots could be useful in slowing down the spread of the virus [173, 328], because of their potential to reduce human to human contacts. We worked with social care workers, as their work practices consists mainly of verbal and non-verbal interaction with care home residents, which were perturbed through contact restrictions and hygiene protocols.

Our study focused specifically on how social care workers use such a robotic system and how they appropriate it into their work practices. We wanted so see how often the robot was used and what for, and what benefits, burdens and obstacles the usage of the robot presented in their daily work routines. Our initial research questions, broadly, had to do with the degree to which robotic systems might or might not be integrated into the day and what consequences their routine use might have for the social care workers and the residents. In order to do so, we established the robot in a care home with no specific instructions as to how it should be used, but with a range of applications available which users could deploy as and when they chose. The clear focus on everyday practice, we suggest, carries with it implications for future research.

5.2 Related Work

5.2.1 Social Robots in Care Settings

Existing research on social robots in care settings has ranged over a number of issues, including the attitudes of care workers towards robots and criteria for their acceptance [213], and indeed of older adults themselves to robots [25, 154, 247]; the ethics of robot use in such settings [319] and legal aspects which pertain [304]. Research has further concerned itself with specific contexts such as dementia care (see e.g. [143, 151, 214]) and disabilities [17, 128]. Others have reviewed the status of robots as assistive technologies in relation to what they might achieve, and what the limitations might be [34, 224]. Although such studies report some beneficial results, as Bemelmans et al. [34] point out, conclusions are seldom robust due to limited methodological reliability, although in recent years the number of studies has grown. Furthermore, in only a limited number of cases has research extended beyond attitudinal studies or short term observations to actually studying the adoption of robots into care over a longer period of time.

Exceptions include Wu et al. and Fiorini et al. [107, 338], both of which entail studies in a living lab context. Further, a limited amount of ethnographic research has taken place involving social robots in residential care settings. The existing studies place more emphasis on the sociality of robots (see e.g. [177, 266, 268]), in one case assessing only conversational elements, with input from an operator, and including care workers in the analysis [268]. The sociality of robots has been more thoroughly investigated in other research areas (i.e., HRI). By way of example, social bonding and intimacy between a person and a social robot have been evaluated by manipulating a robot's affective and other facial expressions in an experiment [177]. Such work, though potentially relevant to the care context, has not specifically involved people from that context. Other work on supporting informal care workers with robots exists, involving long-term co-design activities [214]. Some work has aimed at designing trustworthy care robots [300], but such work has emphasised the direct role of robots as care workers rather than their potential for supporting the work practices of human care workers. Hornecker et al. worked on the triadic interaction between the robot, the care worker and the resident and hence their work constitutes one of the very few multi-actor approaches. They argue that such interaction only works when all parties are taken into account [146]. The need for more practice-based research in relation to robot use has been identified [70, 278], as has the concomitant need for longitudinal studies that focus on the inclusion of social care workers in day-to-day practices as an essential part of human-robot-interactions in care settings. More or less by definition such studies require a qualitative approach.

Studies with a long-term approach either focus on the residents of care homes [70, 133] or focus on older adults living in private homes[27, 125, 196], measuring attitudes and relationship-building with robots. The relatively small number of studies that have actually taken place in care homes (i.e., not in private homes or lab environments) have suggested that social robots have a positive impact on the residents [70] but did not focus on the effects that robots had on the daily work practices of social care workers, even though they demonstrated how social care workers enabled interactions between robots and residents [172] (short term) or how a moderating person is needed [70] the need in principle to integrate with their work practices was not addressed. The study [70] showed that residents enjoy the interaction with the robot and that some intervention in activity was needed. In that instance, however, the interventions were made by a researcher. No attempt was made at that time to identify or analyse how social care workers might intervene the process of using the robot as part of their daily practices. In care settings as such, there is, to our knowledge, limited existing work that primarily focuses on the effects of social robots on the social care workers' workload and in support of their activities in the long-term. One study evaluated effects on social care workers with a NAO robot [139] but did not focus on daily work practices. Another longitudinal study has involved deploying a robot in a care home as an information system to adapt the robot to the information needs of the users, mainly taking a robot-centered perspective [133]. The potential long-term value of social robots has been further discussed by expert users, pointing to the importance of personal value of affection, robot functionality , and social communities forming around robots [92].

There is, then, we suggest, a clear research gap. There is little or no long-term research of a qualitative nature into the way in which robots in practice can support the daily work of social care workers and the obstacles and challenges that social care workers face when making use of a robot in their daily work. Equally, the limited amount of research we detail above suggests that we know too little about issues such as the role of interactional features, of functionality, and of multi-role interactions with a robot within care homes.

5.2.2 Care Workers Attitudes Towards Technology and Robot Usage

Many of the HCI studies that have examined technology use within care settings have focused on technology acceptance [58, 98, 213, 313]. In contrast, Burstein et al. focused less on acceptance and more on awareness of different technologies [58]. They found that awareness about different technologies by care workers is limited, while at the same time there is an interest in using them. Kramer argues that there is an information gap [37], such that care workers are open towards technology and assistive robots [213] and think of them as something useful but mostly know little about them and their possible uses. Accordingly, better communication with care workers about technology opportunities is needed, so that appropriate preparatory measures can be taken for low-threshold appropriation [58].

One of the main arguments against the use of robot technology in care is that care work is complex. Particularly in the area of social robotics, there are concerns that the systems are not robust enough and thus not secure in various respects [34]. Fears regarding loss of employment for care workers are often mentioned, along with ethical and security concerns, [33, 66] According to a survey by Zöllick et al., assistance systems that provide physical support to care workers are generally well received [351]. If the care technology systems are available and provided, they are also used [298, 351], though why that might be is not entirely clear. Other studies have also shown that if technology is developed within the ecosystem of care and developed with participatory design approaches to the specific needs, the technology can be well received and used [216, 327]. We therefore see a need for more research on the practical use of social robots by social care workers aiming at a better understanding of the realities of social interaction with robots.

We contribute to the HCI community by doing qualitative long-term field work and providing insights into the actual interactions between social care workers and social robots, exploring how they make use of a robot in their daily work with residents and how they appropriate it.

5.3 Methods

5.3.1 Study Design

To explore how social care workers use a robot to assist their daily work, we gave them a previously developed robotic system with a view to gaining holistic insights into the interactional quality resulting from robot deployment in the daily work of social care workers with residents.

The study was conducted in three stages. First, we conducted a pre-study that included interviews, the diaries of residents and site visits to explore the use to which the robot was to be put. Second, we embarked on the main study, giving social care workers the robot for a period of 14 weeks. Six weeks were planned at the beginning but the care homes asked us to prolong the deployment. This phase included semi-structured interviews, log-files of the robot, structured note-taking and workshops assessing degree and manner of appropriation (see 5.3.4). Third, applications were tailored to the needs of the care home based on ideas that were expressed by the care workers (see 5.4.2). The methods are described in greater detail in Sect. 5.3.5.

The researchers came twice per week, usually for half a day, to observe the interaction. These meetings were arranged in advance with the care home. Only one researcher at a time was allowed, due to COVID-19 regulations.

The data was collected between October 12th, 2020 and January 12th, 2021. Because this was during COVID-19, we followed strict hygiene rules and created a hygiene concept to disinfect the robot.

5.3.2 Participants and Setting

The study was conducted in a care home within a medium sized city in a European country. The facility was a former monastery and has been converted into a modern care home for 49 care residents. The study participants are a group of social care workers, social service management and nursing management. We worked closely with nine participants, listed in Table 5.1. Despite their different formal work roles, they work together and share tasks. We also consider the residents as participants of this study, as they were an integral part of the interaction and were important contributors to our work. The residents are a heterogeneous group that are probably representative of other care homes as well, though one specificity is that some (approx. 10–15%) of the residents are under 50 years old.

The social care workers provided is an integral part of work within care homes in the German health system[1] and differs from the job role of caregivers. They assist and support the residents in organizing their daily lives in the community and in dealing with personal matters. The aim is to prevent loneliness, depression, and to mitigate loss of health and the increasing need for care. They organize and coordinate individual and group activities, for example in the form of sports, creative or cultural activities. The maintenance of the well-being of the residents is promoted through

[1] In the local context, this group of workers are also referred to as *sozialer Dienst*, which could be also translated as social service.

active support and a wide range of activities. The social service workers were not the only ones using the robot, since usage was open to everyone. Nevertheless, the study showed us that the robot was mostly used by the social service workers as the nursing caregivers are involved in other work tasks. Throughout the paper, we refer to the people working in the care home as social care workers and make the quotes identifiable through the participant code that can be seen in Table 5.1.

Table 5.1 The participants' codes as referred to in this paper, their age, sex, and position in the care home

Participant Code	Age	Sex	Position	Education
SC1	19	F	Social Care Workers	A-Level School
SCM2	54	F	Social Care Worker (Management)	3 Years Apprenticeship
SC3	36	F	Social Care Worker	1 Year Training
GMCH4	52	M	General Manager Care Home	University Degree
CSM5	38	F	Nursing Service Manager	3 Years Apprenticeship
SC6	45	F	Social Care Worker	3 Years Apprenticeship
SC7	74	F	Social Care Worker	1 Year Training
SC8	57	F	Social Care Worker	1 Year Training
SC9	19	F	Social Care Worker	A-Level School

5.3.3 Robotic System and Software

For this study, the humanoid robot Pepper[2] was used. Additionally, the social care workers received a smartphone with an application, which we developed, and which enabled the participants to remotely control the movements of the robot and had a voice interface. The voice interface made it possible to type words or sentences in the smartphone that were repeated by the robot via voice. Further, we equipped the robot with a mobile router, in order to connect the smartphone to the robot and to receive the exact timestamps for the log-files. Besides the hardware described, we also provided the participants with printed instruction manuals. The manual was a modified version of the Softbank manual, with fewer pages and a reduced amount of information. The equipment given to the participants is shown in Fig. 5.2.

[2] https://www.softbankrobotics.com/emea/en/pepper

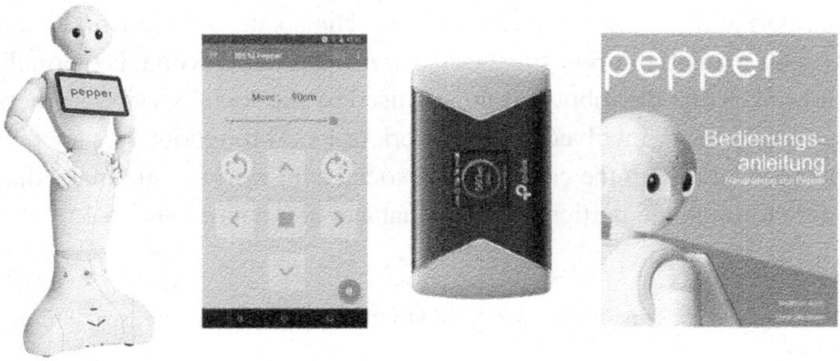

Fig. 5.2 Set up provided for the participants—Pepper Robot, Smartphone with App for remote control, portable router and printed instruction manuals. (Source of pictures, from left to right: (1) softbankrobotics.com (2) tp-link.com (3) own (4) modified, based on softbankrobotics.com)

Most of the applications used (34) have been the subject of previous studies [70, 315] and were reused for the work we present, with the exception of the newly (4) developed robot applications (see 5.4.2). The applications on the robot address three main types of activation: (1) *physical activation*, (2) *cognitive activation* and (3) *psycho-social activation*. In this study, 38 apps (34 old + 4 new) were available to the participants, briefly categorized in Table 5.2. During the study presented here, additional applications were developed and integrated in collaboration with the participants. More details on the newly developed applications is described in 5.4.2.

Table 5.2 Overview of applications provided for the study

Activation type	Description	Example	Total
Physical Activation	Exercises and dances that encourage movement, relaxation and strength training.	Mobility dance	10
Cognitive Activation	Exercises and games that activate thinking and learning processes.	Memory	11
Psycho-social Activation	Applications that evoke meaningful memories or promote social interactions so that mood and well-being are enhanced.	Songs	17

For the period of the study, the robot was stored in a supply room on the 1st floor of the care home, to which social care workers had unrestricted access. The room itself was near the public areas.

5.3.4 Appropriation

In order to familiarize the social care workers of the study with the robot, we combined three elements: (1) a manual, (2) workshops, and (3) support during the usage period. (1) The manual included a description of the functionalities of the robot and explained all technical details, like how to charge the robot or how to turn off the robot. It was there to help the participants in times of uncertainty about the technology and was seen as an instrument of appropriation that did not need additional assistance. (2) The workshop served to explain the robot to the social care workers and marked the beginning of the study. It comprised the presentation of the robot and its basic functionality (e.g., how to move the robot or how to use the touch sensors) and safety measures. Further, the individual applications of the robot were explained and demonstrated. The social care workers had the opportunity to familiarize themselves with the robot and at the same time clarify any ambiguities directly with the researchers. (3) Support during the period of usage was regular and was combined with the data collection such as during periods of observation, which took place on average twice a week. In the beginning of the study, the support was characterized by answers to many questions by the participants. As the study progressed, the researchers took more of a backseat and their role became more observational.

5.3.5 Data Collection, Analysis and Ethical Evaluation

The data collection methods included semi-structured observation protocols, log-files of usage time and usage patterns on the robot, audio files of semi-structured interviews with social care workers and photos.

The interviews were conducted either with one participant or in groups, depending on individual preferences. The interviews took place at the care home and were audio recorded and later transcribed. Observation protocols were prepared in advance and were semi-structured around the topics of: context of use, role of social care workers, difficulties, appropriation and interaction. The data was analysed with a reflexive thematic analysis approach [49] with the MaxQDA software by inductively clustering different topics from the interviews and the notes of the observations in order to identify the patterns that we present below.

In addition to the qualitative data, quantitative data was collected on the integration of the robot into work processes. The robotic system was programmed to create log-files with time stamps to gather insights about usage times. As soon as an app was being used on the robot, the app's name, date and time stamp was recorded. To have the correct time stamp the robot was connected to a mobile router that was connected to the internet. Whenever the robot was not connected to the router, a fill date (01.01.2014) was saved in the log-files, this happened on three occasions, to correct it we compared the log-files to the observation protocol and reconstruct the dates and the usage time.

Ethical approval for the study and its data collection methods was given by the ethical committee of the university (name concealed for review) with the reference number: ER06/2020.

5.4 Findings

Overall the robot was well received and we observed that the social care workers used it regularly in their daily tasks. The findings we report on concern the usage of the robot, the appropriation of it, the attitudes of the social care workers towards the robot and both the perceived burdens and benefits that resulted.

5.4.1 Usage Time and Usage Distribution of the Robot

The recorded usage of the log-file was between October 12th and Dec 4th 2020. Because the study started mid October, the 42.1% of usage time in October is the usage time out of 19 potential usage days. There were 29 potential usage days in November instead of 30, as the robot was being programmed on one day by the researchers. In November, the participants used the robots on 48.3% of the days, the highest usage rate in the study period. Even though the care workers asked for a prolongation of the study until January, there were only four usage days in December. Care workers were increasingly busy with Christmas activities (such as organizing and carrying out an outdoor Christmas market), and there were also times where social care workers were isolated due to COVID-19. Researchers were also increasingly unavailable to enter care homes in December because of a two-week isolation after being in contact with another researcher who was tested positive and because of increasing COVID-19 restrictions (see also discussion in 5.5.5). The median values of the usage over time indicate that the robot was used for about 1 hour and 11 minutes per usage day. This time was calculated by using the first and the last time of the logger, during one day. In this time, on average, one social care worker used the robot to visit four to six residents in their rooms, but group activities also took place as can be seen in Fig. 5.4. From the observations, it is clear that the additional time for starting up, possible delays due to error messages and clearing away the robot is estimated at 15 to 20 minutes, so that a standard gross time of 1:30 h can be assumed.

Table 5.3 Usage Time of the Robot per Month and in Total

	October	November	December	Total
Possible Usage Days	19	29	4	52
Actual Usage Days	8	14	2	24
Usage %	42.1	48.3	50.0	46.2
Median	01:10:12	01:11:05	03:44:27	01:11:05

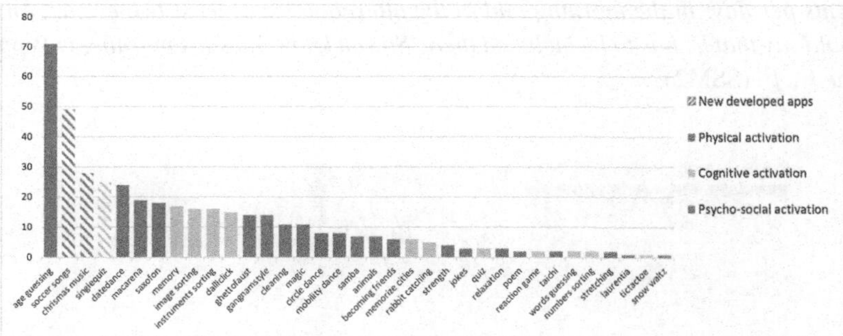

Fig. 5.3 Total usage distribution of the Application of the Robot

In terms of distribution of applications, 67% of the apps used belong to the psycho-social category, 26% to the cognitive category, and only 7% apps to the physical category. With regard to the overall course of the study in Fig. 5.3, it is noticeable that it was primarily the newly developed apps (see 5.4.2) for the facility that have been best received and thus most intensively used, including *christmas music* and *national soccer* (both psycho-social). In contrast, it is noticeable that some apps were hardly used, such as *stretching* (physical) or *tictactoe* (cognitive). The *age guessing* app was used the most, especially from care workers to introduce other care workers and residents to the robot (and to introduce the robot to them). It was perceived as socially stimulating and funny, and also serving as an ice breaker.

5.4.2 Appropriation of Robot Usage

5.4.2.1 Integration of the Robot into Work Processes

The robot was available to the social care workers throughout the course of the study. It was used rather experimentally in the first two weeks (October). During this time, the social service management (SSM2) made sure that each social care worker had the chance to try out the the robot. However, there was no clear pattern to this, the social care workers used the robot whenever they wanted but did not integrate it in the fixed structure of activities that was planned by the care home.

This opportunistic usage of the robot began to change in the following month (November) and became considerably more structured. The social service management (SSM2) created a monthly plan to schedule events at the care home (two events per day), and the robot became integrated into this plan as can be seen in Fig. 5.4. The work plan was initiated and carried out by the care home in order to integrate the robot into the daily work: "*[...] in the weekly schedule we always propose two events per day: in the morning and in the afternoon, so at least two events. And I would say that he has to be included there. So you know Tuesday morning is Pepper time [...]*" (SSM2).

Fig. 5.4 Two pictures illustrating the activities with the robots. The left picture shows an activity plan for the month of November. One activity is listed per day. Among other activities (e.g., movement with music or worship services), Pepper is is listed, including "Pepper visits living areas & the rooms", "dancing and sitting with Pepper", and "puzzle fun with Pepper". The right picture shows two daily schedules with planned activities on one paper each, pinned on a wooden board in the hallway visible for residents, visitors and employees. The schedules include "Pepper visits living areas" on Monday, 9 November 2021 in the morning; and "games and fun with Pepper" on Tuesday, 10 November 2020 in the afternoon

While the management encouraged the social care workers to use the robot during the study, the social care workers could nevertheless decide for themselves. There was no usage obligation. The robot was not only used during the sessions that were announced on the event board but also on occasions when the social care workers visited the residents in their rooms to do the daily check-ins, to talk and provide stimulation for the residents.

The interviews and observation helped us to understand this usage patterns and showed that some social care workers became power users and used the robot more frequently (SC9, SC3, SCM2) while others were reluctant to use it and did not use it often. The power users proceeded to help other social care workers, if there were technical problems or questions about how to use it.

5.4.2.2 Robot Application Adaptation

During the course of the study, there was an increased desire for additional applications for the robot. The social care workers had specific ideas on what the robot should be able to do and asked us to develop these applications. We discussed with the social care workers which of the ideas were important to them to support daily tasks and possibilities were discussed in relation to the capabilities of of the robot and our own ability to deliver such applications within a short time frame. In the end we developed four additional applications, which are shown in Table 5.4.

Table 5.4 Overview of developed applications for care home

Name Application	Description Application
National Soccer Songs	Robot sings the national soccer club anthem and dances
Christmas Songs	Robot sings and dances different Christmas songs
City Quiz	Robots guides through a 10 questions quiz, asking questions about the city where the care home is located
Christmas Quiz	Robot guides through a 10 question quiz, asking questions in regard of Christmas

The applications were well received as one social care worker points out: *[...] I still remember the soccer songs, the club anthems [...] I have specifically one person in mind—absolute Bayern [FC Bayern München] fan. The following day, the man was still happy.* (SC8). The applications were used frequently as can be seen in Table 5.3 and discussed in Sect. 5.4.1.

5.4.2.3 Attitudes towards Interaction with the Robot

As a social care worker said, the employees' attitudes can be classified as *"positive throughout the study"* (SC8). Because the robot had not been used regularly in the facility before, Pepper was seen as something new: *"We found it exciting to deal with it. We were pleased because we were allowed to work more with it, because it was something different than just reading something out, [...] you could try something new yourself."* (SC9) In addition to the effects on the people working directly with the robot, there were also effects on the social care workers who did not work directly with it: *"They [other social care workers] pet Pepper on the head, talk to it or want to take pictures. So that's like a little celebrity driving through the hallways. I think they just kind of get excited then. [...]. Then they talk about it for a short while. It just lightens the mood."* (SC9).

Despite the generally positive response, there were also a few cases among colleagues in which the robot use was viewed critically. *"[...] they just said it directly, that robots don't belong in here. That the robot could not replace human relationships. [...] He simply could not imagine that the robot would interact with him, because he simply said that a robot cannot replace a human. [...] That was his purely ethical opinion."* (SC6).

It was noticeable that the people with the critical voices seldom involved themselves in interactions with the robot. At the same time, we did see that sceptical social care workers who briefly got involved in the interaction with the robot tended to change their opinion. They would meet the robot on the hallway and try out applications such "age guessing" or they would observe residents' reactions.

The general manager of the care home saw effects on the team. He believed that the social care workers had begun to see the robot not only as an animation tool but partly as a team member: *"Ultimately, I can also confirm that somehow he [the robot] is part of it [the team] now. He is somehow humanoid, [so that] it has something of a team member. [...] We will miss something."* (GMCH4). The manager looked back positively over the course of the study and the impressions gained. Through the study, "*negative fantasies*" and "*fear of contact with the topic of robotics could be reduced for the time being*" in the entire facility, so that Pepper "*created much more openness*" among the employees (GMCH4).

5.4.3 Burden of Using a Robot

Although relieving the burden of work is often a main argument for promoting robots in care homes, difficulties arose during the study that could not be overcome by the social care workers and were thus perceived as a burden to them. Because the social care workers focused on one-on-one care, the robot was primarily used in the residents' rooms. These were not always located next to each other, which resulted in relatively long walking distances for the workers. Here, the robot was mainly pushed by hand and only rarely controlled remotely with the smartphone app, as the control with the smartphone was slower than pushing the robot by hand. The pushing was perceived as physically demanding: *"The transport, I find it actually very difficult. Because you get the feeling that it [i.e., the robot] is a bit unstable. [...] I have a small doorstep and [...] I couldn't push it over there. It was really difficult [...] for him to get over this threshold, and I mean, he's not light either"* (CSM5).

In addition to the physical burdens, the social care workers described some mental stress. They were often afraid of breaking something, as the robot was seen as an expensive piece of kit. Some social care workers were also concerned about possible accidents and what to prioritise in case of accident: *"What if the resident chokes on something while the robot is heading into something. Where should I act first?"* (SC1). They also described other kinds of mental stress. For example during the regular delays that occurred when starting up the robot: *"[...] if I know there is a technical problem and I can't get it solved, I don't have a solution. That makes me feel anxious."* (SSM2). This indicates the need for more transparent communication of errors in terms of why and how long these occur, and how to fix them. As the robot became part of fixed scheduled activities in the weekly plan, the error-proneness and lack of transparency also led to organizational stress: *"As I said, when it doesn't work, it's a bit of a burden, because you're so internally prepared to use it now or want to use it. And when it doesn't work, you first have the technical problem. [....] And at that moment you also have the problem that you can't implement the plan you have for today"* (SSM2).

Besides these obstacles, there were also minor difficulties in the operating process itself. The social care workers tended to focus on individual care, so that many bedridden residents were in contact with the robot. Due to the height of the beds and the height of the robot, social care workers found it difficult to position the robot well. It was necessary to position the robot close to the bed during games so that residents could interact via a touch-pad, but at the same time it was important to ensure that the robot had enough space during dance performances. Due to the lack of sensors on the center of the robot, where the "hips" are, the robot often did not recognize the bed, so the robot arms became caught on the bed. According to a social care worker, it would be desirable that when the robot *"gets closer to the bed, maybe the motor system can then turn off"* (SC6).

5.4.4 Work Relief through Usage of Robotic System

Despite the obstacles to usage described, it is clear from the data that the robot was perceived overall as a support to social care workers in their everyday work. Sometimes, the presence of the robot itself was enough to positively stimulate the residents' well-being: *"So just the fact that Pepper was present. That's a highlight for the residents. Yeah, it was something different. My face, they see that every day. And there was something coming, [...] something out of this everyday austerity. So I personally found that a relief. I come in with Pepper and the resident is already happy. Goal achieved"* (SC8). In addition, the social care workers argued that they were able to take on a more passive role, since the robot itself became the main actor: *"[...] that you are not the main actor, but you can take a very small step back and just accompany the whole thing. [...] the main idea comes from Pepper"* (SSM2). This could mainly be observed in longer interactions but sometimes the robot was also used for shorter periods, for example we could observe that the robot was sometimes placed in the residents' rooms to wake them up from their nap, by playing music.

One particular scenario stands out from the observations, which exemplifies the potential relief provided by a robot. A resident regularly complained to the social care workers that she wanted to go home. The resident has dementia and had an episode in which she was convinced that she was 40 years old with a family waiting for her at home. Two social care workers used the robot to distract the resident: They introduced the resident to the robot and tried to find commonalities between the her and the robot: *"Well, first of all, we found an introduction to say that he's almost a bit like E.T.: She knew him now, too. And the attention was immediately bound by this first sentence and her memory of this film. And then we played Dalli Dalli [famous game show from the 70s and 80s] and the music alone from this Dalli Dalli, which has accompanied people for years, sometimes decades, was a real door opener. [...] She was able to get completely involved with the robot during that time and she was suddenly clear of this great distress, at least for the time that this was going on"* (SSM2). While the social care workers cared for the resident with the robot, there was an emotional relief for all involved. The robot was a vehicle for the social care workers to steer the conversation in a different direction and there was an uplift in the resident's well- being for this short period of time.

This scenario shows that in some situations it is not enough to just place a robot in front of a resident. It needs moderation by an experienced social care worker in such cases to encourage interaction between the resident and the robot. In addition, this scenario shows that social care workers need to know the life history of the residents in order to create a suitable connection between the robot and the resident.

Even if, contrary to what was originally intended, the social care workers did not see any new time resources devolving from the use of the robot, this activity was perceived as having an affective benefit, since the robot as the main actor positively influenced the situation.

5.5 Discussion

The findings revealed that in accordance with our research questions, social care workers readily incorporated the robot as tool to cope with daily-work practices. In the following we will discuss what that means for sustainable integration and the appropriation of the robot. Our contribution demonstrates the way in which positive outcomes of robot use are closely tied to practical organizational matters.

5.5.1 Sustainable Work Process Integration

Sustainable integration requires us to take into account internal and external actions. By 'internal', we mean the adjustments that social care workers and residents need to make at various times. By 'external', we mean those factors which support those adjustments, such as training, technical support, and so on. Our findings suggest both factors to be important.

A robot can be especially intrusive both at a physical and a symbolic level, as other work has suggested [280, 295, 302]. Therefore, sustainable integration requires a holistic approach to tackling the needs of residents along with those of different kinds of social care workers. In the latter case, it is particularly important to recognise the varied roles that social care workers might have. Because social care workers have different roles and responsibilities, it is crucial to work out a clear strategy with regards to the robot, who might be deploying it at any given moment, and what uses it is to be put to. As multiple social care workers may be working with the same residents, there is an evident need for a holistic coordination.

The social care workers defined three components as part of their appropriation of the robot's functionalities. (1) This included getting to know the robot. At this point, willingness was crucial. Most social care workers were proactively interested in using the robot, even those who were not officially participating in the study. (2) At the same time, the management also encouraged the social care workers to use the robot, which might have added to their curiosity and their motivation to engage. (3) However, not all social care workers felt comfortable incorporating the robot in their work practices. The value of the robot and its functionalities, failing full automation, relies entirely on the attitudes of workers to it and, more importantly, their ability to integrate it into their care routines. The delivery of good care in and through robotic applications, then, depends on the practices of social care workers, in accordance with Schmidt's argument, that practice entails both routine and normative alignment [272].

Once past this initial familiarisation stage, social care workers began asking for new apps, encountering some technical problems and progressively widening their focus over time as other needs took priority. This was, for instance, the explanation for the flattening of usage over the Christmas period. For sustainable usage, we would argue, it is important to recognize the need for systematic and proactive planning. Different social care workers may have different needs and desires for the robot usage and it is important to recognise them and to find ways to best integrate them. In our study, a work plan and an outline of tasks for which the robot might be used was created by the workers themselves. This was a dynamic process, however, as social care workers, having got to know the robot and intuiting what kinds of things it would be capable of doing, expressed wishes in terms of the apps that might be developed, and, as we saw above, these apps turned out—perhaps unsurprisingly, as other research had similar experiences [41, 193]- to be those most used.

5.5.2 Overcoming Uncertainties

As matters progressed, the social care workers reported on their fears and the mental stress experienced when using the robot. These were exacerbated when technical issues arose. Reducing these fears requires supportive measures [308]. On the part of the researchers or providers of the robot, it is necessary to provide low-threshold technical support to reduce these obstacles. While social care workers used the opportunity to ask researchers for technical support whenever they were present to collect data, they were much less likely to do so via email or by calling the researchers. It is difficult to know exactly why, but we feel that a more immediate response facility, perhaps embedded in the robot itself, including a support phone number on the back of the robot, might prompt more response. Whether or not it prompts increased use of help facilities, it would at least increase the workers' sense that reliable intervention is at hand.

When support was asked for, the request mainly came from "power users" in the institution. These were also the providers of help to other less knowledgeable social care workers, a phenomenon well-attested in a number of domains [241, 344, 346]. For these users, additional training seems helpful as it would enable them to provide internal support and become experts themselves. Certain people are inevitably more motivated or better at certain skills than others. There is a 'social distribution of expertise' [253] that necessitates a careful approach to training and dissemination but also to identify such users quickly in order to give them the needed support.

Given the generally positive view of management towards the robot- even suggesting that it had become a "team member"- there is then a need to carefully allocate roles and responsibilities. There was a need for clear communication and support structures when things went wrong as it was difficult for some social care workers to understand why the robot had technical problems. Workers were also concerned about breaking the robot, again largely a matter of inexperience and something previously encountered [31, 39]. To overcome this, the social care workers were told that the robot was insured.

5.5.3 Interaction Design of Usage

The usage data shows that the robot was used regularly in the first two months. The applications show a clear trend towards use of the entertainment of care home residents, more than use of applications that are more health oriented like cognitive games or physical activation.

In regard to interaction design the robot was evidently seen as more useful when performing functions that can best be thought of as socially interactive and stimulating. This involved two levels of social interaction: either shared activities between a resident and a robot via music or games, or stimulating interactions between one or more residents and a social care worker. (1): The most popular apps involved music, dance, and other playful activities (see Sect. 5.4.1). These apps offered social interaction between a person and a robot and they were used as such (i.e., by singing or dancing together, or by playing a game). As also found in previous research in the care context [70], a social care worker typically acted as a moderator between a resident and the robot in such scenarios, even if the robot was primarily used in group settings, with one robot and several residents interacting. In the study presented here, we observed that there was a preference on the part of social care workers to use the robot with single residents, mainly in their rooms. Typically the social care workers would bring the robot to the room, start the robot and help the resident to interact with it. It may be that the COVID-19 pandemic was a partial explanation for this, in that there may have been a desire for reduced group activity, but we believe that this was not the only reason as some group activities still took place. (2): We could also observe that the robot was used as a social facilitator, fostering communication and interaction between residents (sometimes several) and a social care worker. This supports conclusions from previous work which emphasizes the role of social robots in fostering and even generating interaction [92]. In the care home, both residents and social care workers were speaking with each other about the robot and they were using it together. We believe that this (along with the moderator role of the social care workers) is also related to the specific role of *social* care workers who were using the robot, and who would not leave a resident alone with any manipulable tool. This also resonates with the results of Kidd who observed that care workers typically enabled the interaction of residents with social robots [172].

Earlier work had already shown that, in the context of a care home, interaction between residents and the robot was only possible with the active involvement of a social care worker [70, 139]. The main difference in this study was the way in which social care workers were attentive to the demands of their daily work and assessed the value of the robot in relation to the degree to which the quality of their work was improved, rather than it simply being made easier. We had assumed in the beginning that they would use it to occupy some residents with the robot, without their involvement, so they could complete other tasks in parallel. This turned out not to be the case. Analysing our data, we see a variety of reasons for this: (1) The robot usage was seen as a positively enjoyable thing and therefore they preferred to be part of it. (2) As the development of the robot software was made together with social care workers, they felt themselves to be integral to successful implementation. (3) Social care workers knew they were observed, and may possibly have been subject to a certain 'observer effect', though we have no reason to believe such an effect persisted beyond the initial period.

Social care workers, while appreciating the fact that the robot occupied the residents, also felt that these socially interactive applications had a positive effect on their well-being (e.g. *"I come in with Pepper and the resident is already happy. Goal achieved"*—SC8).

5.5.4 Usage Preferences

When we take a closer look at the usage of applications two things become apparent. First, the 'age guessing' application (older development) in particular was well received and second, the applications that had been specifically produced for the care home ranked amongst the most successful ones. The 'age guessing' application, as the name suggests, guesses the age of the person standing in front of it, based on an algorithm that measures features of the person's face. It was frequently used by the social care workers to introduce the robot to other coworkers and residents and has proven to be something of a 'warm up' application to get people to further engage with the robot, something that has also been seen in other studies [291].

The frequent usage of the newly developed applications was, we believe, for three interconnected reasons. For one, the care workers have told us that they felt that, if the robot was to be used, it was to be used 'properly' (Management of Social Service said to social care workers: *"remember that Pepper is not going to stay forever, if you want to use it, do it now"*—SSM2). Second, the apps were used as a way of managing the risk of spread of the virus. Specifically, two of the three requested application were singing applications (soccer songs; christmas music), requiring active involvement on the part of residents (but sometimes were also used to wake up residents from their nap). During the pandemic, however, singing was discouraged as it was seen as a potential vector for the spread of the virus [14]. Since that was not recommended, they let the robot sing and tried to animate the residents to sing with the robot. That had the result that the social care workers did not have to sing themselves, since they were seen as potential vectors for introducing the virus. Third, the applications were asked for by the social care workers based on their perceptions of how much activity on the part of residents might be generated. From our observations, we know that the residents enjoyed interacting with the new applications.

A flattening of usage occurred in the third month of our study, despite the fact that social care workers wanted to prolong the study (including the availability of the robot). A flattening of usage would in the first instance point to a possible novelty effect wearing off [188]. However, we found that several other issues around Christmas and COVID-19 played a role in using the robot less than in November (see presentation of the results in Sect. 5.4.1): (1) In December, major Christmas activities were conducted, such as a setting up and holding of an outdoor Christmas market (this normally takes place within the city but was cancelled due to COVID-19 regulations). With rain and snow, the conditions were not ideal for using a robot. The Christmas activities required significant resources, especially from the social care workers. As shown in our results, the robot was not used as a stand-alone artefact, but always with a social care worker. At this time, social care workers were involved in seasonal Christmas activities and therefore had less time for the robot. (2) The work plan that had involved activities with the robot was designed for the month of November, and given the changing schedule around Christmas and the resources needed (in particular social care workers), no activities were planned in the work plan. The robot was free to use for all social care workers but not scheduled upfront. Bringing us to the conclusion that the planing of the management to use the robot and to include the usage in the timetables of the social care worker had substantial effect on the usage time. (3) Additionally, COVID-restrictions were becoming increasingly strict around Christmas. It became difficult to enter the care home for the researchers, reducing the possibility to help with technical problems if these occurred. Based on the three listed conditions we believe that the decrease of usage was not because the novelty of the robot wore off, but because the robot, in quite specific circumstances, was no longer a good 'fit' for the routine work of the season.

5.5.5 Reflections on Appropriation in Regard to Daily-Work Habits

Our intention was to establish how and whether the robot could become a tool that was perceived as useful in the daily work of social care workers in the care home. To do that we observed how the care workers appropriated [131] the device for their work. This implies that the robot needed to be adapted to the work practices of the social care workers and some consideration of whether adaptation to the robot implied some changes in their work was necessary.

This study mixed two concepts of appropriation that are described by Stevens and Pipek [131] as 'End-User Development' [191] and 'Tailorability' [339]. The End-User Development aspect, giving the users the ability to modify the software without coding, came through the smartphone application that was given to the social care workers. The smartphone gave the social care workers the ability to control the robot and to build small programs (sequences of different applications and making the robot speak). The tailorability aspect, adapting the software to local needs, was introduced through the participatory design [325] of additional applications that were specifically adapted to the needs of the care home. While the smartphone application (to control the robot with the smartphone) was not often used and described as "buggy" and complicated, the tailored software was more successful, the workers were proud of the applications they came up with and the software solved a problem for them (singing in times of COVID-19). This example aligns with prior participatory design studies for robots [186, 280] and shows that even if the tailoring of software on the robot with participatory design methods creates additional work, it results in a meaningful usage and identification with the development.

What became quite apparent even at the outset, and consistent with previous research [70, 139], is that the interaction with the robot needs assistance, with individuals playing discrete roles. It is not possible to leave residents alone with the robot, social care workers are needed to guide the interaction and therefore have to adapt their work practices to the robot, if the robot is to play a consistent role in the care home.

5.5.6 Work Relief Effects

The care sector, and social care workers in particular, are under an increasing pressure to meet work requirements [167]. Technology in general [212] and robots [160] more specifically are sometimes seen as having the potential to relieve work pressure. Our study showed, in contrast, that the new tool (robot) increased choices with regard to the kind of care that could be provided but it also added to workload rather than subtracted from it. The robot augmented quality of life for residents in various ways but there was almost inevitably an overhead for social care workers. The potential of the robot lay primarily in its contribution to an increased job satisfaction or a change of job roles as more technical competences are needed and therefore a wider spectrum of abilities are needed by the social care workers. Even so, one might consider the 'relieving' effects of the robot less in terms of time or task burden, and more in terms of alternative, and positive, possibilities, as was for example demonstrated by the case of the person with dementia who got distracted by the robot because she connected it to E.T.

Another aspect of work relief is the one of preparation. Social care workers usually prepare the interactions they have with residents and think of activities they will engage in with the residents. The robot, because it was pre-programmed with various activities, meant that they were able to spend less time on the detailed planning of activity. Robots, in this context, are not proven time-savers but, on our evidence, do enhance possibilities.

5.6 Lessons Learned

Our findings suggest a number of relevant factors that can be instructive for future development. They make it clear that, as an HCI audience should recognise, successful integration will require a great deal more than technical functionalities geared to delivering services to residents. The lessons learned are presented in the following list.

1. **Development with Local Workers:** The most used applications were those which had been requested by care workers which, while not surprising, nevertheless testifies to the principle that it is naive to assume that the needs and

desires of both patients and care workers are self-evident. Care workers know the needs of residents better than outsiders. More importantly, they are much better able to judge feasibility.

Rationale: The results in Sect. 5.4.2.2 showed us that local workers have contextual, or situated, knowledge that relates specifically to individual resident needs and of the constraints embedded in the facility's routines. An example which illustrates this was the development of an application which reflected an interest in football, and which seemed to have a direct impact on the wellbeing of one resident. The request for such an application came from SC9, who recognised its potential value.

2. **Integration with Ongoing Activities:** Feasibility is largely a function of recognising existing structures, roles and responsibilities, timelines and routines. In our case, the robot was integrated into the weekly schedule so that the residents and care workers were jointly aware of when it would be used, and what for. The process of integration, we argue, should not be enforced hierarchically, but should be designed by the facility itself.

   Rationale: As can be seen in 5.4.2.1 the social care worker typically has a work plan of activities that is offered to the residents. SSM2, for instance, included the robot as integral to scheduled activities, with specific social care workers using the robot almost daily.

3. **Involvement of Stakeholders:** As care is complex and multiple professions are working with the same residents it is important to involve them in the process of bringing in the robot. There is an evident need for all parties to be heard and informed about the process.

   Rationale: As discussed in Sect. 5.5.1 the robot can be intrusive at a physical and a symbolic level. The observations of the human-robot interaction showed us that the robot needs a holistic approach to be sustainable integrated into the different facets of the work in a care home which is filled with people from different professions and different backgrounds.

4. **Voluntariness:** With regard to the use of the robot, the principle of voluntariness seems to be important. The value of the robot in practical operation depended substantially on it being recognized as useful and meaningful by the individual worker, and on the enthusiasm, willingness and commitment to use the technology on the part of the individual was critical to its appropriation.

Rationale: Sect. 5.4.2.1 and 5.4.2.3 showed us that not all social care workers equally enjoyed working with the robot. Some become power users (SC3, SC9 and SCM2) and others just used it occasionally. In turn, it would appear that the degree of enthusiasm demonstrated by some workers testifies to the need for voluntary commitment.

5. **Technical Support:** While technical support will always be necessary, the degree to which it is, the degree to which it is needed depends on the willingness of care workers to manage problems, to recognise that technical errors can happen and they are not normally at fault and should not be made accountable. Such matters are as much organisational as individual. They rely on a health recognition by all parties that problems can arise.

   Rationale: As the social care worker (SSM2) reported in Sect. 5.4.3 the usage of the robot did also result in mental stress as it was unclear how technical issues can be fixed and this would potentially ruin the planned activity. Most technical issues are quick to solve and the help of technical support can ease that process.

6. **Power Users:** We saw that some care workers stepped up as power users, pushing the development, helping co-workers and taking responsibility if problems arose. A degree of 'enskilling' would enhance the social care giver's role and equally enhance the prospective role of the robot.

   Rationale: Sect. 5.4.2.1 showed us that some social care workers became power users. At the same time we can see in Sect. 5.4.3 that some are stressed out by the possibility of technical errors. We believe the power users could help with the majority of technical issues, if trained accordingly.

7. **Quality of Care instead of Workload Reduction:** Care workers perceived the robot as providing useful adjuncts to their everyday work tasks. Nevertheless, there was little benefit in terms of personnel or time resources. Rather, benefits have more to do with new possibilities of delivering an enhanced quality of care.

   Rationale: As SC8 and SSM2 point out in Sect. 5.4.4 the robot helped them in two ways. For one it helped to offer new possibilities for interaction. The robot was something new and helped to distract from otherwise repetitive activities, an effect that might wear off with time. SSM2 added to this that the robot has a set of applications, a repertoire that enhances the possibilities, with familiarity, for the social care workers.

The lessons learned here are applicable for the system we present (see 5.3.3). Much may change as robots themselves acquire new functions and as software evolves.

5.7 Limitations

We also point to some limitations around this research. Through the regular visits within a long-term study, the researchers built up a certain relationship and degree of trust with the participants. Some positive findings could thus have been influenced by the Hawthorne effect, though we have no reason to believe this was the case [257]. The residents are highly vulnerable to the COVID-19 pandemic that occurred during the study, and activities at the facility had been generally reduced. Group settings were largely to be avoided. The pandemic had an influence on the study and would have included more visits of the researchers, if conditions were different. Since the number of participants remained small, the results can only be applied to other care homes with reservations. Consequently, the findings are valid to the setting described here and are not necessarily representative, though we would argue they provide a starting point for future research on care home work.

5.8 Conclusion

A number of themes emerge from our work on the daily work practices of social care workers with a social robot. Firstly, we could show that social care workers do make use of the robot in their daily work and perceive it as a useful tool. Nevertheless, there is little support in our work for the view that robots offer significant opportunities to save time and reduce workload. However, because the robot was mainly used for social interactional purposes, it gave social care workers new possibilities that broadened the spectrum of options for handling care situations and on occasion reduced the amount of preparation required to provide residents with meaningful interactional experiences. This resulted in an enhancement of care quality as the primary driver of use.

Secondly, our work indicates the need for a careful consideration of the 'social distribution of expertise' (see e.g. [253]). Some social care workers are more skilled, or more motivated, than others. It was clear from the outset that a degree of initial acceptance was critical for the integration of the robot. That general willingness translated into a number of suggestions concerning what kinds of application might best serve their work practices. It is perhaps not surprising that those applications, when implemented, turned out to be the most often used and it does testify to the importance of recognising the local knowledge and skills that go into doing social care work. Tailoring the software through participatory design methods made the integration of the robot into routine care practices much more straightforward, since it established not only what kind of applications might prove interesting and useful but also information about when, where and how such applications might be deployed.

In our view, much of the hype around the role of the robot in care settings is a product of some techno-centric assumptions and of very limited research into the settings into which robots will be placed. Put simply, the vision that robots will significantly reduce the human workforce, for the moment, remains just a vision. Our research shows that the contingencies which arise in the setting are important. The effectiveness of the robot depends on decisions about what tasks it might perform, when it might be asked to perform them, what human skills might be required in order that it consistently delivers them and, arguably most importantly, what coordinative work will be required. The latter, in particular, requires attention to what skills are needed on the part of social care workers, how they are to be trained, and what kinds of application actually provide support for them in their work.

In conclusion we argue that social robots like the one we used are ready to be used within care homes but their successful deployment depends on an understanding of the routine work being done in the care home. Further it is important to adapt the robotic systems to local needs, the robots do not work ‘out of the box’. The work entails a level of cooperation, and coordination of activity, and a level of competence that is not universal. It seems like it is not the readiness of the technology nor the willingness of care homes that constitute the main obstacle to development. Rather, what is needed is a rigorous understanding of the ways in which the use of a social robot can be successfully and sustainably integrated with daily work As much as anything else, that entails a rigorous understanding of what that work actually is.

5.9 Acknowledgement

We would like to thank all involved parties. This work is a shared result and we are deeply thankful for the given opportunity. We were given the chance to work on this study during a difficult time and are grateful to have been welcome at the care home.

The presented work has received funding from the European Union H2020 Programme under grant agreement no. 101016453.

Video Connecting Families and Social Robots: From Ideas to Practices Putting Technology to Work

6

This chapter was published in the Universal Access in the Information Society Journal as *Schwaninger, I., Carros, F., Weiss, A., Wulf, V. and Fitzpatrick, G., (2022). Video connecting families and social robots: from ideas to practices putting technology to work. (2021). Universal Access in the Information Society, 1–13.* It corresponds to the P3 in the List of Publications.

Abstract. *Technology use is a socially embedded process, especially when it comes to older adults and care. However, the restrictions associated with the COVID-19 pandemic have limited social contact to protect vulnerable groups in care homes, and even if technology use has increased in other areas, there is little known about the potential uptake of communication technology and changes in social interaction in the care context during a lasting crisis. This paper explores changes in communication technology use triggered by the pandemic at two care homes, using a qualitative diary study, online interviews and observations, and in-situ interviews within the care home with residents and workers. Our findings point to increasing use of tablets and video conference software triggered by COVID-related experiences, with implications for living and working in care homes. We also characterise the isolation experience of the residents, the workers' concerns about the residents and changes in social interaction. We observed new areas of technology usage, associated changing work practices, technical affinity issues and context-specific attitudes towards future technologies. While the pandemic has triggered the use of communication technology in care homes on a small scale, this has also caused increasing workload and in particular articulation work, which requires support structures and the re-definition of work roles.*

F. Carros, *Design, Development and Sensemaking of Human-Robot Interaction in Care Settings*, https://doi.org/10.1007/978-3-658-45233-9_6

6.1 Introduction and Background

Technology use is a socially embedded process. When it comes to older adults and care contexts, practices of technology usage are both social and collaborative [251], and are often embedded in care networks [279]. Informal and formal care [198, 251, 306], inter-generational relationships [201] and help networks [99] are important for the use and uptake of collaborative (care) systems. The appropriation and adoption of technology further involves (mainly hidden) configuration work [30, 216, 293].

The care context in central Europe has been however under severe restrictions since the outbreak of the COVID-19 pandemic in March 2020, which caused an increase in physical distancing on the one hand and higher dependency on digital tools on the other. Older adults and other high-risk groups especially have been protected[1]. As a consequence, care homes were closed, visits were prohibited, and resident mobility was restricted. This resulted in a loss of family and intergenerational face-to-face contact [201], and for some even a lack of contact with fellow residents and personnel.

While digital tools may provide an opportunity to promote communication [114] and connection [262] and even to reduce isolation/loneliness [105], residents in care homes do not always see digital tools as an option for stimulation [114] on a physical, cognitive or emotional level. Despite an increased overall tendency to conduct social and economic activities [83, 201] and even research virtually [202], there are yet unequal opportunities to mitigate the risk of social isolation [262], and digital accessibility issues are known from previous COVID research [184]. This engenders the risk of double exclusion in a society that relies heavily on communication technology (CT), namely from physical contact and from digital participation [282]. Further, while COVID-related restrictions have been ordered top-down by governments, earlier studies have shown how healthy ageing is strongly perceived as an active achievement that includes the management of lifestyles, health and illness, and the active balancing of social life and financial and material circumstances [287]. Economic resources and digital skills are crucial factors to reduce risks associated with the pandemic [262], i.e., for social, cognitive and physical stimulation. Both residents and people working in the care sector however often lack the required

[1] On March 11th, Dr Hans Henri P. Kluge, WHO Regional Director for Europe has announced the pandemic: "*COVID-19 impacts the elderly and those with pre-existing health conditions most severely. In a spirit of solidarity, we all have to be ready to contribute our part to protect those people at highest risk. As individuals, practising good hygiene and prevention measures as well as applying measures of social distancing, including avoiding crowded places, continue to be very important.*" [337]

digital skills [114]; while this problem is not new [91], it has become more pressing when physical social contact is no longer an option.

Special circumstances like crisis situations may reveal societal norms and understandings in a very concentrated way. Further, restrictions may trigger developments and change habits to cope with special circumstances, also depending on the ideas that have been developed and 'lying around' [110]. Assistive technology has been developed and discussed to support older adults since many years [135] (e.g., including safety systems, security, monitoring, communication, and entertainment systems, and home automation [135, 312]). The use of technology for communication in particular can be a response to crisis situations [132], and research suggests its potential as a response to isolation in the care context that has occurred since the pandemic [114]. However, there is still little known about the actual use of CT at places like care homes during a lasting crisis. We aim to get a better understanding of how the experience of the pandemic has affected attitudes and use of CT, and associated processes of social interaction at these places.

The overview of this article is as follows. In the next section, we present a description of our methodology, including our mixed-method study and the chronological accounts of the data collection in relation to our research process. Subsequently, we present our findings on changing technology usage associated with experiences of the pandemic, including experienced isolation and changing work practices associated with increasing CT usage, and mixed attitudes towards future technology. We follow with a reflection on solutions to old problems exacerbated by COVID and new issues, along with a discussion on readiness for engagement with new technology, and lessons learned from shifting crisis situations. While the pandemic has triggered the use of CT in the care context to some extent, this also requires solutions for new problems associated with increasing workload which we propose to be tackled in future work.

6.2 Methodology

6.2.1 Chronological Account and Context of the Study

At the time of the outbreak of the COVID–19 pandemic, we were preparing a study that involved social robots at care homes. However, it quickly became clear that the pandemic had also affected the care context and especially the social circumstances there. We therefore decided to conduct a study on the experiences of the pandemic and its effects on technology usage and changing life and work practices. While no robots were present at this time, we found that other technology was lying around (to be described later in this section), and we were interested in investigating technology usage and attitudes towards future technologies, such as social robots, as a response to this crisis.

We collected our data in the two care homes at different time phases (see Fig. 6.1). Phase 1 was the strictest phase concerning the restrictions in Western Europe where our study took place, and it included most of the first lockdown (March-May 2020). In this phase, the care homes did not let other people enter, which also resulted in families being torn apart. We used this phase for method preparation and recruitment. Restrictions to receiving visitors were lifted rather quickly after the first lockdown and they were not reinstated with the same strictness later. In Phase 2 (May—June 2020), visits were allowed under severe restrictions, including window visits[2], or container visits[3]. A limited number of visitors per day were allowed, enhanced hygiene rules like wearing masks, and rapid COVID tests were in place. In Phase 3 (July 2020—Jan. 2021), visits were possible under lighter restrictions. We collected different types of data in Phase 2 and Phase 3 in the two care homes.

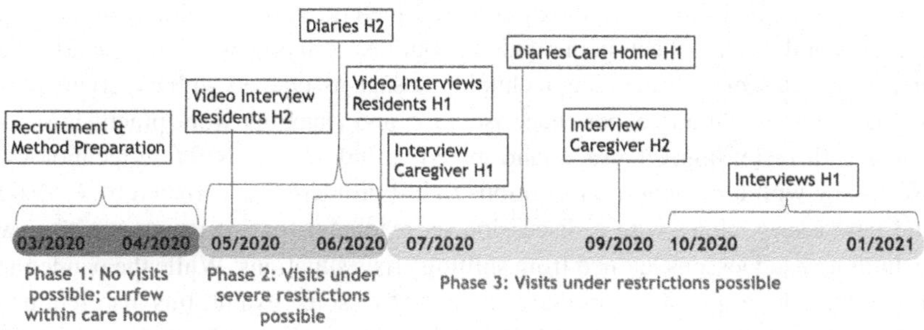

Fig. 6.1 A chronological account of the restrictions and data collection phases

6.2.2 Research Setting and Participants

The two care homes that participated in the study are situated in a German-speaking country in Europe. The care homes are anonymized in this paper and they are referred to as H1 and H2. H2 is within a big city, while H1 is situated in a medium sized city. H2 is a long term partner of the university of one of the co-authors and had a regular exchange with the researchers until the pandemic. Over a period of the last

[2] i.e., visitors communicated with the residents through a window without entering the care home

[3] i.e., containers were installed in front of care homes especially for visits

four years, the care home worked together and experimented with different digital tools like tablets, smart assistants and social robots. H1 has been also cooperating with the university on additional topics, like education.

Both care homes have religious roots and belong to catholic institutions. They are both connected to multiple other organisations within their buildings (H2: Kindergarten, Residential Assisted Living; H1: Monastery, Administration of Organisation). H2 is approximately twice the size of H1 in terms of permanent residents (119:49). H1 is connected to a monastery with nuns living within the care home.

In the study, we had a total of 18 participants: 9 residents and 9 care workers. The residents were all cognitively able. Before conducting the study, we acquired ethical approval from the university. The residents were invited by the care workers, and their participation was based on their willingness and ability to participate. The residents were between 55 and 93 years old, 4 of them male and 5 female. An overview of the participants is presented in Table 6.1.

Table 6.1 The participants' names as referred to in this paper, their age, gender, care home facility (pseudonymised) and role in the respective care homes

Name	Age	Gen.	Facility	Role
R1	55	m	H2	Resident
R2	86	f	H2	Resident
R3	85	f	H2	Resident
R4	85	m	H2	Resident
R5	84	f	H1	Resident
R6	60	f	H1	Resident
R7	86	m	H1	Resident
R8	60	m	H1	Resident
R9	93	f	H1	Resident
C1	62	m	H2	Social Service
C2	54	f	H1	Management Social Service
C3	19	f	H1	Social Service Intern
C4	36	f	H1	Social Service
C5	52	m	H1	Management Care Home
C6	38	f	H1	Management Care Workers
C7	45	f	H1	Social Service
C8	74	f	H1	Social Service
C9	57	f	H1	Social Service

6.2.3 Data Collection and Analysis

We conducted a diary study and online interviews with 18 participants (see Table 6.1. and Fig. 6.1 for a timeline of methodology) in the two care homes[4]. We collected data from 9 residents by conducting a diary study and interviews (phone or video interviews), and from 9 care workers by conducting face to face interviews.

We created paper-based diaries for residents to fill out for four weeks per participant, with one entry per day. The diaries included open questions and playful activities (such as story completion and postcards (see e.g. Fig. 6.4) to be sent to researchers and a robot) to gather insights on the role of technology in day-to-day activities, on the experience of social contact and the pandemic, and on perspectives on the next generation of CT (i.e., robots), as we were interested in perspectives on the role of (both current and future) technologies triggered by the pandemic. The diaries were filled out by nine residents at two care homes (i.e., four and five, where the imbalance of numbers has been a result of the voluntary participation). To complement the long-term diary study, the participants who volunteered to fill out diaries were invited to participate in remote interviews with us using a chosen communication medium (e.g., telephone or video telephone using a videoconferencing software like Skype). Four interviews were conducted with residents of H2 via telephone, and five interviews were conducted with residents of H1 via video telephone, with the assistance of a care worker. To gather insights about care workers' perspectives, we also conducted interviews with nine people working with residents, one from H2 and eight from H1 (based on the workers willingness to participate).

We recorded the interviews using recording software for audio/video. We transcribed the audios and videos and we coded them along with the diaries. The data was analysed with a reflexive thematic analysis approach [49] with the MaxQDA software by inductively clustering different topics. We generated, reviewed and defined themes to interpret results that we present below.

We collected data out of interest in the effects on the experiences of the pandemic on technology use, and therefore did not actively change the environment with

[4] Note, we did not intend for a comparison between the two care homes. However, we had several reasons for inviting two institutions to participate in this study: (1) We were in the middle of a collaboration with the two institutions at the time of the outbreak of the pandemic, (2) we wanted to get a broader view on the topic, (3) the situation and topic was delicate and it required a highly exploratory approach, and (4) only four residents volunteered to participate in our virtual study in H2 who we had invited first. Methodologically, the only difference in approaching the institutions was the slightly different time of data collection (as shown in Fig. 6.1). Furthermore, as described earlier, we conducted telephone interviews with the residents in H2 and online video interviews with residents in H1 responding to their preferred medium.

research by asking participants to use technology (this was also not possible because of the restrictions, at least in the beginning of our study). However, both institutions had access to technical devices, even if in slightly different ways: Within the 6-years cooperation with the university of one of the co-authors, H2 had received around ten tablets and the residents were trained to use these devices by the university until the pandemic started. H1 on the other hand was not a long-term partner, however they received tablets from a national telecommunication company in Phase 1 and they did not receive any training.

6.3 Findings

Across our data, we identified effects of the pandemic on the care homes, new technology use, and attitudes towards future technology as reported by care workers and residents. In the following, we present findings on experienced isolation, workers' concerns about residents, and changes in social interaction. This is followed by technology use at care homes, including the use for social interaction and physical activation, associated changing work practices, technical affinity issues, and subsequently, attitudes towards future technology.

6.3.1 Effects of the Pandemic on the Care Homes

Experienced Isolation

Especially in the first phase, visits were forbidden ("*visits = prohibited!*" (R2)), where later, containers were installed in front of care homes dedicated to visits, as shown in Fig. 6.2). The restricted visits was difficult for the residents: "*Of course, the […] people first had to get used to the fact that their people—their relatives—were no longer allowed in the house. That was, I would think, the biggest setback.*" (C2).

Besides the restrictions on receiving external visitors, the isolation of residents was also enforced through restrictions within the care homes. Group activities were reduced: "*Yes, our sports activities and bingo and what we all do. That also has to be cancelled and we can only do it with so much distance you just sit with two men […] so that we don't get too close.*" (C1). Furthermore, it became normal that everybody had to wear a face mask within the care homes at all times to protect others from possible infections. These hygiene rules were intended to slow down a possible infection but also resulted in limiting the residents in understanding other people—since their sense of hearing is quite often limited, they depended on being able to use several senses to compensate: *"They had to keep the minimum distance, they had to keep the mask and I would say, the old person can't hear so well, he can't see so well. And he depends on it, to see the mouth move and also that you can get close to speak loudly and so on, and all that was not possible. That made some residents feel insecure, rather than comfortable, because you couldn't reach the other person, so to speak."* (C2).

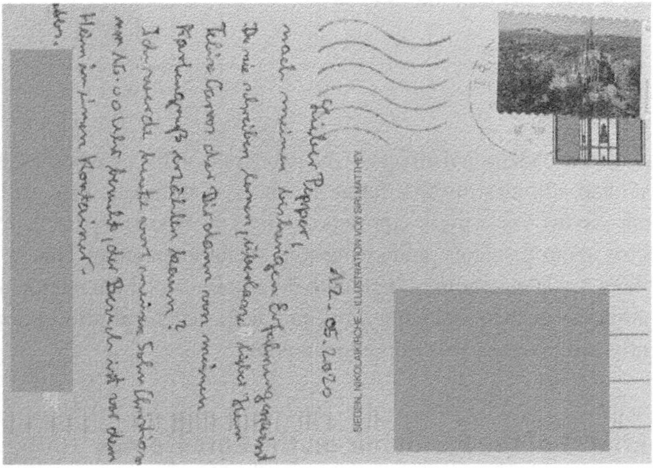

Fig. 6.2 A postcard a resident has written to the researchers describing the visits in containers: *"[...] I will be visited by my Son Christian today at 4 p.m., the visit will be in front of the home in a container [...]"*

The rules also had the consequence that the residents did not receive physical contact when it was not necessary. Before COVID-19, it was normal that a care worker would give them physical contact, like patting the shoulder. Also, visitors normally provided physical contact via hugging or holding hands. All of this was restricted and took away an embodied sensual way of communicating without words.

As the residents are quite a heterogeneous group, not all of them experienced isolation in the same way and they had different perspectives on loneliness. One resident reported that: *"sometimes I need the loneliness"* (R2), suggesting that loneliness has positive sides for her. However, this view was not shared by all residents, as another person told us: *"I want my kids to be allowed to visit me again"* (R3).

Workers' Concerns about the Residents

Care workers reported safety concerns regarding the residents. On top of the usual health-related risks (e.g., "*the usual influenza will come additionally in November [...], then we are likely to have our noro-virus in the care home [...], as every winter*" (C1)), workers were concerned about not being tested for COVID-19 in Phase 2 (which changed later): "*great that we are not being tested here. Everyone could be a spreader even without symptoms*" (C1). However, there was also awareness that the residents had gone through several other personal and global crises in their lives, where the pandemic may not be perceived as difficult by some of the residents. One resident compared the experience to that of being a prisoner: "*I am 94 now. Four years of war captivity in Siberia, I survived it all. Now I am locked in like a prisoner. I need to die anyway, I want to be able to move around freely during the last months that I have. Within the remaining months I have left, I expect to be able to move freely." (C1).*

The care workers recognised that increasing dependency of people as they got older challenged their sense of self-confidence, and the workers would do what they could to try to promote self-confidence: "*in my view, this is the most important, because these people that are here, they all have been like us, [...] they were able to do everything like we do and now all they can do is to experience deficiency [...]. I think the most important for them is that they are able to do it on their own, completely and especially to understand on their own. This will promote their self-confidence and a sense of mastery.*" (C2).

Changes in Social Interaction & Communication

Both residents and care workers reported changes in social interaction and communication because of the pandemic. In the care homes, there were fewer group activities happening, and in the beginning of the outbreak they had to eat their food alone in their room ("*lunch is being brought in our room, you know, we don't go into the big dining area downstairs, and in the morning, a plate with breakfast is put here and in the evening as well*" (R1)). This resulted in more monotonous daily routines, and residents wanted to have more activities and see people ("*of course, then there would be more variety*" (R5)). Residents also made efforts to keep physical distance ("*of course I avoid all the people*" (R2)).

The interaction between residents and care workers changed, where the face masks had an effect on communication. For residents who were depending on facial expressions a lot, the workers "*lose their personality, when wearing this thing, in my view, some of the personality is getting lost [...][my fear is that] everything is getting more mechanical, [...]one is nothing but a number*" (R4). Because of the restrictions, there was also a decrease in face to face interaction with people from outside the care home. Residents were however missing their family members ("*I wish that my children are able to come and visit me again*" (R1)).

6.3.2 Use of Communication Technology

Using Communication Technology for Social Interaction and Physical Activation

In response to the restrictions and associated experiences, off-the-shelf technology usage was initiated by care workers at both care homes. These were also used because of fewer visits and to increase safety.

To communicate with people from outside the care homes, both institutions were using video telephone from different corporate technology companies ("*we have used WhatsApp a lot, a lot*" (C1), "*a friend usually comes regularly to visit me, now at times of COVID-19 [we have been using] Skype*" (R1)). However, a care worker also stated that video conferencing platforms were used only by approximately 10% of the residents during the pandemic, compared to 0-1% before the pandemic: "*Those who have been using it appreciated it [i.e., video telephone]. Among over 100 people however, if only ten people were using it, this is not a very good quota, because [compared to]feeling closeness for real, [...], sitting next to a person, [...]*" (C1). However, among the people who were motivated to use communication software, a care worker said: "*We really took our time for every resident who wanted to be able to use it on their own. Skype is always a bit of an obstacle, but WhatsApp, everyone has it, so we showed it to everyone, even the relatives, how they could use it, so the relatives can contact the residents on their own or the other way around*" (C2).

Video telephone was appreciated a lot, as care workers tried to "*realize everything as good as possible, to take into account every opportunity and guidelines*" (C2). A care home reported how they got support from a national telecommunication provider with tablets and apps such as Skype or WhatApp which made this even easier (C2), and according to the care home, the use of video "*has worked out very well*" (C2). The use of video telephone was also perceived useful for the people who were able to use it. As visits became possible with special measurements, if they "*could compare between Skype and original visits, they preferred Skype, because they said: 'What should I sit here, where I can't get close to them and so on, then this is better, I can also hear better etc. when doing it over Skype*!" (C2).

Video conferencing tools were also used by care workers to conduct exercises with residents ("*we continued with our program—movement with music. And we had rehabilitation sport twice a week, and we did the same for movements with music*"

(C2)). However, this required a lot of space, "*we had the big advantage that we are very spacious here and we could separate people from each other*" (C2). Workers made use of the videoconference tools with screens: "*with bad image quality and bad audio quality, we used the TV in the cafeteria and in the rooms to transfer and show our sports program*" (C2). People were able to gather and still keep the minimum distance from each other. This way, "*they could still see their usual caregiver on TV*" (C2), and another caregiver could still come by to the people and support them physically.

However, the predominant use of tablets for communication required assistance by care workers or relatives. The lack of visits had an effect: "*It [video telephoning] is installed on my laptop. When my son comes, he will help me*" (R2). The help of relatives was missing, which required care workers to assist. With the use of tablets, on the other hand, residents also had more "*access to media*" (C2) to access online information. While residents had been less open to use technology for this purpose before, "*where nothing else was possible, they might have accessed it and say: 'Okay, now I'll give it a try'. This was actually very nice.*" (C2).

Changing Work Practices

The use of technology at the workplace entailed changing work practices, with both positive experiences among workers and challenges, and issues related to digital literacy, accessibility and additional workload.

Digital literacy was an issue in general. While care workers needed to assist residents in using communication software, they were also in need of guidance: "*This is currently difficult because of Corona, [however] if there were seminars, also for the residents or small videos with explanations[...]*" (C3). However, following the engagement with videoconference software to enable residents to communicate with people from outside the care homes, the workers started to use video telephone as a new habit in their private lives: "*This [video] telephoning has become a habit at some point. We also used Skype, everyone who had the opportunity to set it up on their phone*" (C2).

The use of CT also increased the workload. Workers assisted residents with setting up conference software: "*We enabled everyone who had the option to have Skype—with their phone. We guided everyone [...] if they weren't familiar with it yet*" (C2). Because "*most residents are not able to see very well, so with this, let's say, video conference with the outside as one could say, they do need a little help*" (C2). In addition, there was also considerable work involved in setting up appointments for the calls: "*people [i.e., family members] called at least a day before Skype calls should happen*" (C2), and "*some [residents] even booked appointments for the entire week*" (C2) to speak to their relatives or friends. During the virtual conversations with relatives, care workers also reported that they were often directly involved e.g., to support the communication and give short updates to relatives (e.g., regarding the status of the resident).

This increase in workload came on top of an already high general workload, exacerbated by restrictions of other people visiting the residents who would normally help. On the other hand, the care workers also found it rewarding to use of technology. They experienced learning they were able to support the residents' needs ("*This was actually great*" (C2)).

Technical Affinity of Care Workers

As care workers were forced to use CT at their work and because they have used video telephone frequently ("*we have used WhatsApp a lot, a lot*" (C1), see also Fig. 6.3), their technical affinity appeared to increase. While care workers had not expected to need technology for their work before the pandemic (C2), "*once that nothing else was possible, they might have have accessed the technology, 'okay, now I'll give it a try'. This was actually nice. Also the fact that we have to engage with it – what is possible, what is good for everyone*" (C2).

Fig. 6.3 An advertisement at H2 to communicate with relatives via CTs like Skype and WhatsApp

One care worker expected the new habits to remain at the care homes after the pandemic. Videotelephone "*[...] will not be gone only because of opening steps, instead, this [i.e., videotelephoning] is an additional option now [...] Of course, Skype will play a role in the future, the same goes with WhatsApp calls with video.*

This will not just disappear because of the opening, but it is becoming an additional option" (C2).

However, while they accepted that the use of technology would persist, they were also aware of their skills gap. The generally low technical affinity of people working at care institutions (i.e., care workers and people in management positions) was mentioned (C1). Technology usage would mean additional workload. A care worker however stated that there were too little offers from the government to promote digital literacy (C1).

6.3.3 Attitudes Towards Future Technology

As we also asked questions about attitudes towards robots (i.e., our intentional focus before the pandemic had started), we found mixed attitudes from residents. Some residents were sceptical about the potential for robots: *"Modern technology can open new perspectives but can never replace human affection!"* (R4). The resident sees technology as not smart: *"Computers don't think by themselves. Everything has to be inputted beforehand."* (R4). This opinion is not shared by everyone though as can be seen in a comment that was made about a robot from previous experiences at H2: *"Yes, I like him. He is not too tall, he doesn't frighten me and I want to have him with me"* (R3). A resident who had experiences with Pepper wrote postcards to the robot, saying *"I would love that you can pronounce my name [...]"*, and *"[...] tomorrow Pepper should come and visit me"* (see Fig. 6.4). One resident who had previous experiences with Pepper even reported that his family was interested in the robot: *"My kids want to get to know Pepper."* (R2).

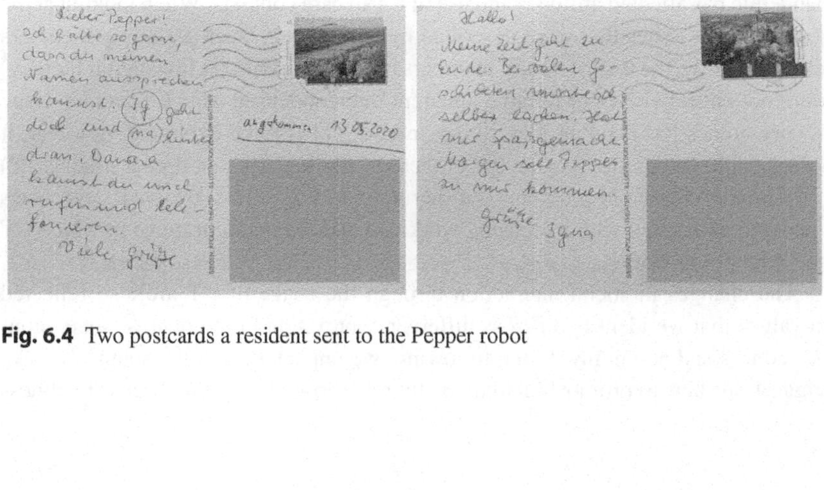

Fig. 6.4 Two postcards a resident sent to the Pepper robot

A care worker on the other hand expected a social robot to be problematic for residents because of their very feeling of deficiency: *"They are likely to build up a relationship right away [...] 'oh he is cute', 'he is so pretty' [...] I believe they wish for too much, even if unconsciously. And if Pepper [i.e., the robot] does not react to what I would be able to provide, a person who is used to bear a lot of deficit may feel deficient again. [...][T]hey will think immediately, 'is it my fault', 'why does he [Pepper] not understand me'..."* (C2).

The residents and care workers had many ideas regarding the functionalities of a robot and the tasks it could do for them. One care worker saw it as something that would be helpful to give orientation to people with dementia: *"For the elderly, demented it could be something to give them orientation. [...] Maybe saying things like: 'At noon we have this and this as lunch' or 'This morning at 10:30 the Bingo game will happen'"* (C2). The residents saw several tasks that a robot could do like cleaning or bringing things from one place to another. A resident who had previous experience with Pepper had a specific idea on how he would cooperate with a robot: *"I am taller than him, I will work on the things that have to be done on the wall and he can help me on the ground."* (R3). A resident also thought the robot could measure the temperature of people (R2). One care worker however expected the robot to promote physical contact: *"it is rather perhaps even contact-promoting or also motivates, rather to the participation in groups. And that could be counterproductive, of course, regarding Corona"* (C1).

6.4 Discussion

The aim of our study was to get a better understanding of how the experience of the pandemic has affected attitudes and use of CT in care contexts. When looking at the results, many issues that were described by participants (such as family members living far away, the loss of family members because of death, or low digital literacy) were not entirely new but became apparent in a condensed way. CT and its usage at care homes has been an idea 'lying around' [110] to be exploited, however, it took a pandemic to increase the use of these technologies (even if on a small scale) to respond to these old problems. Furthermore, as CT use has evolved, new issues appeared associated with increasing workload and the configurations in which CT has been put to work.

The changes in social interaction through the usage of CT are also reflected in values that we identified across different contexts and time phases: social connectedness and autonomy. In the following, we unpack these values and we make suggestions how to promote them in the future. Subsequently, we discuss readiness to engage with CT, followed by lessons learned from shifting crisis situations, where we pull out implications for living and working in the care context where technology is now expected to stay: the need for the acknowledgement of increasing workload and support structures on several levels, and the re-definition of work roles and processes. Furthermore, we discuss an outlook taking into account the next generation of CTs, followed by methodological remarks and limitations.

6.4.1 Values in Social Interaction

Social Connectedness

The restrictions of the pandemic [337] have further reinforced the notion of care homes being isolated in society. As a group prone to higher risks, older adults living in care homes have been experiencing isolation at times during the pandemic, with fewer visits and increased restrictions in mobility at times that have affected social interaction.

Experiences of loneliness or solitude were mentioned not only in relation to the restrictions. They were also associated with losses of relatives or partners that residents have experienced previously (i.e., before the pandemic) or with relatives living further away (not caused by the pandemic), as also discussed in the literature [100, 105, 222, 288]. Further, experiences of loneliness were positively and negatively connoted, where it is important to take into account that loneliness does not simply mean the absence of people but it has been also described as the absence of context and connectedness (which is not necessarily provided by the presence of people, as one can also feel lonely in groups) [88]. As residents were open to new forms of social interaction and communication that has also involved technology, it is important to aim for connectedness and context in a meaningful way, also taking into account specific experiences of loneliness.

There were accessibility issues with face masks for residents who were dependent on reading facial expressions during face-to-face communication. It is therefore important to design hygiene rules in a way to enable communication and connectedness where possible, especially given the impact these rules had on social interaction and communication.

CT was introduced and used for social interaction. It took a pandemic to discover the potential of these devices in the care context for certain tasks. Care workers also reported positive experiences of using tablets in their work, as they described the use of video conferencing tools to promote a sense of community. Residents who have been able to use video telephony to communicate with relatives appreciated it, where the social potential for digital technology for older adults in general has been already highlighted in previous work [134]. Advantages were even mentioned compared to face-to-face visits if these visits were restricted: people were able to hear/understand better as compared to visits without being able to touch each other or the wearing of masks. However, digital competence is a requirement and digital literacy issues excluded a lot of people (a known barrier for technology use among older adults [108] also in previous COVID research in the care context [114]), or they required the help of care workers to make contact with people outside care homes. We therefore see digital literacy a key requirement to promote social connectedness with digital tools. On the other hand, promoting digital literacy may be also used to establish social connectedness, e.g., by creating social activities around digital literacy promotion.

Autonomy

Using CT has the potential to promote residents' autonomy in daily life (e.g., through access to additional information and / or communication). However, as shown in previous work [147], autonomy is a multi-faceted concept [61] and sometimes one facet can be impaired to ensure another one. While some residents interacted with relatives virtually, these modalities of communication also came with new dependencies due to the additional articulation work [273] (to be further described in Sect. 6.4.3): Residents required care workers to assist them in setting up and actually being in contact with people from outside care homes, and schedules were made in advance. This articulation work may involve choices to be made for residents to promote their autonomy, for example, in when, how and with who to communicate.

Autonomy is also tied up with mobility and the regulation of the environment with regards to mobility, where restrictions forced people to have fewer social contacts and to move around less freely. Our data was collected in settings and with residents with little mobility (i.e., compared to the rest of the population), and residents also expressed that they were not happy with these sorts of restrictions when they felt they had little time left. Therefore, the restrictions in autonomy (i.e., not being able to move freely) may be perceived in a particular way at older age, especially as people living at care homes may not have the same choice of following or breaking such rules autonomously.

6.4.2 Readiness to Engage with CT

The pandemic has stimulated residents and care workers to deal with the impact of the pandemic by trying out new forms of social interaction and communication, especially with the use of CT like tablets. Care workers also experimented with video conference tools to conduct physical activities and support people. At some point, residents were also open to try out the use of tablets for information retrieval and to communicate with relatives.

However, digital communication practices have varied among older adults even long before the pandemic [35, 166], which is also reflected in our data showing only a small group of residents have actually used these new devices. Here, it would be possible to make these residents internal facilitators for other residents to help them with the use of tablets, as it has been the case in other research (albeit with active older adults not living in care homes) [279]. This could also be used to meet known barriers among older people such as asking for assistance with technology usage [108].

However, readiness to engage with technology that is lying around depended on various factors: Obviously, a crucial factor was the availability of tools lying around. Further, the willingness of care workers to support residents in digital communication was also crucial and required them to actively engage with technology. A lack of digital competence was also mentioned in one care home, tied to a lack of support by the government, social security and in care workers' (ongoing) education. Care workers needed to adapt their work practice to the new forms of support they were giving to the residents and by doing so changed their work habits from helping family members to meet the residents within the care home to helping the residents and the families to use this CT. Readiness to engage with new technology could be promoted through the availability of devices, training and other support structures, institutional (work) culture, working conditions and the roles of workers. This has implications on relationships and the coordination of care networks, which connects to previous research in the care and healthcare context [130, 207, 320].

6.4.3 Lessons Learned from shifting Crisis Situations

Crisis situations can change imaginaries of what is possible to respond to old and new problems. Digital transformation has been accelerated in various sectors during the pandemic [83, 105, 195, 201], and we see changes also in the care context to some extent. Most notable however, the pandemic experience has changed some imaginations of what is possible or what is valued, where solutions for old problems were found. Many relatives were already living far away from residents, resulting in very few visits from them and very little contact. The problem persisted. While technology had been there and relatives had been sometimes far away, the pandemic experience has triggered the usage of CT.

Engagement with new CTs and associated care practices requires an ecosystem perspective with interdependencies at least between politics, organizations, workers and residents. Care workers mentioned political agenda requirements to support the use of technology during work and to include aspects of digital competence into education and ongoing education of care workers. Institutions across care and politics can offer these opportunities and by doing so enhance digital literacy and access to state-of-the art technology. On an organizational level, the willingness of care workers and residents is needed to engage with technology in day-to-day work and life. The care institutions are also the ones who engage in institutional collaborations, for example, with research institutions and private corporations. Such corporations can provide access to participation in development, where participation can provide learning for older adults [186]. The care workers also need a certain level of readiness to engage with new tasks and use technology to fulfill their goals in day-to-day work, provided that they are offered opportunities by care institutions to do so, and that the technology is developed for them to actually support their daily work.

Last but not least, residents need a certain level of digital competence. This connects to previous COVID research which suggests a perspective to understand digital tools as they are being used by residents, not merely as an instrument but also as a learning process that needs professional support, infrastructure and training [114, 248].

The use of new CTs requires a collaborative perspective as they involved a triune of residents, care workers and family members. For the most part, digital devices were not used in a dyadic human-human interaction between a resident and a family member, nor has there been an omniscient and omnipotent agent [273] or user, but they have been used in a triple of residents, care workers and family members. The three parties are needed to establish a connection between each other. The family calls the care worker to make an appointment, the care worker then makes an appointment between the resident and the family. Once this is established, the communication still needs additional support. The care worker sets up the video call and quite often stays in the room with the residents to help out and to give an update to the family. This also means that all three parties have to appropriate a new technology and learn about the changing dynamics that come with it, where they need to articulate the distributed (work) activities [273] to interact and communicate in spite of distance. For the resident, it adds a layer of transparency, since in this situation they are present and talks between the family and the care worker about the resident and their general well-being are done with them instead of outside of their room. As shown by Lahtiranta et al. [182], care workers are building a bridge between the people who are not used to the technology and the technology itself. Furthermore, they are mediating the bridge to the family.

The use of CT adds to the workload which requires recognition, support structures and new work roles. The workload of care workers was increased during the pandemic also because of the three party arrangement. While one could argue that fewer visits may have freed up some time for the workers, health and care sectors may also rely heavily on adult family members' support [201]. Also, care workers have mediated technology use on top of extra pandemic-related workload (e.g., due to changing hygiene concepts and evolving rules that came with the pandemic), which is related to the (mostly hidden) configuration work [30]. Care workers however saw a value in supporting residents to virtually communicate with family members and to use technology for sports and entertainment (even though one could argue that they were running the risk of extra workload in the future). This may have an impact on the organization of work and we see a potential for new work roles, as it is not clear yet who is going to formally take the work that technology brings in. Therefore, roles of care workers need to be re-defined, which may also involve changing job descriptions and specific time allocated. There is also a need for increasing staff and support structures to change work processes (which may also involve union involvement). In order to be sustainable, the work that technology brings in needs recognition and support. Furthermore, even if some older adults may be able to use technology independently after a while (as shown in previous case studies [292]), this may not be always the case given various barriers [108].

6.4.4 Outlook: Implications for the next Generation of CTs

Before the pandemic started, we were in the middle of planning a study with a social robot to be deployed at the care homes. However, at the time of the outbreak in Europe, this has not been possible. It was the idea of engaging with off-the-shelf CT 'lying around' [110] that has been available to be picked up, not social robots.

While CT (i.e., tablets) that has been exploited during the pandemic may be much different from a robot, it is still worth discussing an outlook for future technologies and what can be learned from responses to this crisis for the next generation of CTs. This does not mean that the findings are directly transferable, especially as expectations towards robots are very specific and grounded in pre-conceptions [280]. However, it also became very clear in our study that certain aspects were not a matter of technology itself, but a matter of the context of living and working at care homes in how CT has been integrated into work and life.

First of all, readiness to engage with technology needs to be taken into account. Here, political and institutional requirements are important as well as the role of workers and residents that are not necessarily a matter of technology but a matter of the context. While robots are however not "lying around" usually, a limitation is here that robots may require a different set-up and even more training than already discussed in the case of these much simpler devices, i.e., tablets. The moderator role of care workers as part of a triune that we identified in the use of tablets may be even more important to consider for the use of robots (which also connects to previous work stating that the actions of care workers are crucial for facilitating human-robot interaction in certain care processes [70, 72, 146]). While robots are often envisioned in a dyadic human-robot constellation [146, 277], it is important to acknowledge the dependency of residents on care workers and their key role in configuration work. Our study has shown that CT usage can lead to new dependencies which requires future work to explore opportunities for new work roles or additional staff. While it is often discussed that robots offer more *intuitive* ways of interaction [16, 228, 295, 348], the reasons for this additional workload is also grounded in the task itself (rather than only interaction modalities of a technical artefact, which is nevertheless a key design aspect too). The work practices and ways of day-to-day care along with the crucial role of care workers for putting the communication with relatives to work must be certainly acknowledged also when designing and integrating future generations of CTs in this context, which also needs further exploration in future work.

6.4.5 Methodological Remarks and Limitations

On a methodological level, we aimed to adapt our research process continuously to the special needs and conditions of this time. This research also involved sensitive or difficult topics such as social isolation, experiences of being alone and related issues. Because methodology can affect people especially when they are being confronted with sensitive questions, we took a lot of time, research and preparation ahead to develop material (i.e., the diaries and the interview questions asked). The feedback has been mostly positive as people were able to reflect when filling out a diary, also dealing with emotions. As some care workers needed to assist people to fill out their diary (adding more work), this may have also facilitated a connection (and this very fact may have affected the data to a certain level). The postcards may have also shown how the residents connected with us and we got a better sense of their experiences.

Nevertheless, we were mostly constrained to use virtual data collection methods. While we explored different methods (e.g., diaries, story completion, postcards, and virtual interviews), there may be more suitable methods to gather insights of the field, such as in-situ interviews or ethnographic methods. The assistance by care workers in filling out diaries may have resulted in biases. Furthermore, we discussed the three party arrangement which involves residents, care workers and family members. We focused on the residents and the care workers as a key part of this arrangement. A limitation of our research is that we have not collected any data from the family members with this party as such, and which we propose to do in future work.

Last but not least, the pandemic came abruptly and affected life and work (not only at care homes), which has caused stress that has persisted. One of many other stress factors is the risk of infecting a highly vulnerable group. A limitation of this study may be therefore that the work practices may not be representative, as care workers may have also suffered from the effects of the pandemic in regard of metal and physical health.

6.5 Conclusion

The restrictions in care homes during the pandemic have caused old problems to become more pressing, such as low digital literacy and distance of residents' family members. In addition, new problems like experienced isolation, fewer visits and changing interaction and communication practices have been associated directly with the pandemic. The pandemic has triggered increasing technology use to respond to old and new problems. This has also entailed changing work practices of care workers and increasing workload. Because it is likely that technology is there to stay, future work needs to tackle the changing work and communication practices associated with technology usage at care homes.

We see implications like the need to re-define work roles and increased staff, the need for support structures to tackle issues of digital literacy and support for ongoing education and support for teaching skills for care workers to support residents' independence in digital communication. For a more equal access to CT, it is important to provide multi-level support on taking into account different levels of digital literacy among care workers and residents in future work.

Acknowledgement

This research has been funded by the Doctoral College TrustRobots TU Wien, and by the European Union H2020 Programme under grant agreement no. 101016453. Ethical approval for the study and its data collection methods was given by the ethical committee of the University of Siegen with the reference number: ER06/2020. As human participants were involved in the study, they provided informed consent. The datasets generated and analysed during the current study are available from the corresponding author on reasonable request. We want to thank all our participants (especially the care homes, care workers, and care home residents) to contribute to our study.

Citizen Participation in Social Robotics Research

7

This chapter was published in the Book 'Meaningful Futures with Robots—Designing a New Coexistence' as *Carros, F., Langendorf, J., Randall, D., Wieching, R., Wulf, V. (2022). Citizen Participation in Social Robotics Research. 206–215*. It corresponds to the P4 in the List of Publications.

7.1 Introduction

If science fiction stories are to be believed, robots will eventually take over the world as artificial intelligence enables a new form of sentient life superior to our own. The truth, however, is much more prosaic. How robots interact with humans now and in the future depends on what they are designed to do, and in turn, the way they are designed will depend on what uses we can put them to. Who gets to decide is of huge importance. Disregarding science fiction, there are other, more realistic scenarios which have great importance for the future of our society. They include such things as the future of the labor force, and with it the financial wellbeing of sections of the population. History is full of examples of designers paying too little attention to, or making errors about, the wants and needs of the public, with consequences which have sometimes been detrimental to society. Developments in weaponry are an obvious example. Another way of putting this is that the design of technology is not simply an engineering problem. It is just as much about the social world and the rights and responsibilities of the people in it. Developing hardware and software which meet the social and other needs of users can be difficult, as users are heterogeneous, their wants and needs vary, and meeting individual needs does not always result in wider social benefits. In the case of social robots, which are the subject of our research, we argue that making sure possible benefits are

F. Carros, *Design, Development and Sensemaking of Human-Robot Interaction in Care Settings*, https://doi.org/10.1007/978-3-658-45233-9_7

maximized means getting together with users and other stakeholders to identify what is desirable and what is possible. This is just one example of what is sometimes called "participatory design," but which is itself an example of "citizen science." Citizen science in general refers to projects where citizens themselves provide or collect data and possibly analyze the results in collaboration with researchers. There are many benefits. Science itself becomes a more democratic process, because more people have a say in the direction that research should take. Citizens can be a valuable resource, saving time and money in research and bringing new insights with them. Citizen science, then, can be a valuable learning tool for both researchers and citizens themselves.

But what is citizen science exactly? In this chapter we give our view, showing how the development of useful social robots can be influenced in beneficial ways by engaging with citizens. In our research we involved citizens in the research process by showing them a robot, demonstrating what it could do and its limitations, and asking them how they might use it and what they would not do with it. The results from these questions influenced the future development of the robot. This gave the development a certain credibility and helped us to balance the different needs people have in relation to social robots. We should emphasize here that social robots are quite different from robots per se. Social robots fulfill a different role than robots have done before. Social robots are developed to interact and work with humans—they do not work in closed factory environments, but react to what humans do and say. They are designed for their ability to interact with human beings, to provide responses to users' general behavior, and to answer questions from users in a way that approximates the way humans interact with each other. The fact that it is the quality of interaction that matters is one reason why citizens, or users, have an important role to play.

7.2 Why Citizen Science

When social robotics research is conducted within real world settings, potential ethical issues quickly become apparent [187, 295]. These may involve, for example, the responsibilities given to a social robot (e.g., whether a robot should remind people to take their medication) or issues around data privacy (e.g., who gets access to the data produced by a robot's camera). In this regard, citizen science can be a helpful tool. Citizens are involved in the research process so that researchers can learn from their experiences and reactions. They often have different perspectives on robotics topics, and may see things that the researchers have not thought of simply because they have a pragmatic, real world approach. They will, we can assume, be able to see what robots can and cannot do, and the potential benefits for themselves and others of involving robots in their day-to-day lives. Again, put simply, this acts as a reality check for researchers who might otherwise have quite mistaken assumptions about users' priorities. Showing and discussing use-cases, e.g., using robots to answer older adults' questions in a care home, helps researchers to understand whether interactions of this kind should be encouraged and developed further.

7.3 Our Own Experience with Citizen Science

Our activities with robots in care settings began in 2017, following on from earlier activities related to implementing technology to enhance the wellbeing of residents in care homes. Our research was carried out with the social robot Pepper, which has a humanoid appearance, and involved working with older adults and caregivers. In 2018, however, we began to see the need for wider discussions about the implications of social robots, and saw an approach based on citizen science as a way forward. These discussions focused on issues such as whether robots should be part of the German care system, and, if so, how they should be used in that context (e.g., the robots' roles and tasks, and how they could cooperate with care workers). The work with the citizens was part of a research communication project that was funded by the German Ministry of Education and Research, and belonged to a campaign that aimed to foster discussion with citizens about the future of work. As part of the project, we organized over 50 events for citizens, and used these as opportunities to collect views on a range of issues.

Our approach was not to use the citizens as co-researchers, as is often the case, for instance, in climate research, e.g., the European Marine LitterWatch project [7], which allows citizens to report litter around bodies of water. We decided against using this approach because we did not see opportunities where citizens at large could collect data themselves. Rather, our aim was to provide insights into the current technological possibilities for and with citizens, and, based on this, to jointly discuss and develop ideas on how the robot could be used in future care scenarios. This was important for us because social robots may well become a new element in future care, and in this sensitive area there should, we believe, be a debate throughout society about whether and how this new element might be used in future care settings. In this way, it became possible for us to exchange ideas about use scenarios with citizens who are not experts in the field.

We ultimately tested and combined three different formats for our citizen research:

- 1. Lectures from experts
- 2. Discussions with citizens
- 3. Participatory design workshops

All three approaches were enriching and helped us to gain new insights. Our final method, combining the three formats, was organized as follows:

- 1. Presentation of the robot
- 2. Presentation of use-cases in care settings
- 3. Brainstorming on use-cases in care settings; team building
- 4. Group work: design workshop
- 5. Presentation and conclusion

At the beginning of each event, we gave a presentation on the current state of our research. This introduced participants to the technology used and the uses of these prototypes in real-world contexts (see Fig. 7.1). We then answered initial questions from the participants, such as "How long does the battery last?," or "Can it cook food?" We did this to create a context in which to introduce the opportunities for, and barriers to, this kind of research, and to provide a more realistic view than the one portrayed in the media.

In the case of the design workshops, citizens worked on care robotics concepts in small groups, and the task of the participants was to develop usage concepts based on interactions around a specific task (e.g., the scenario of responding to an emergency alarm, envisaging how such a situation might evolve). The goal was to find desirable applications for the robotic care technology presented, and to describe these by means of text and image flow (see the example in 7.2), and thus to integrate citizens into the design process.

These concepts were then presented to the researchers and other participants and discussed with the group. A diverse range of real-life issues came to the fore, and questions of data protection and ethics were also discussed. The results of the design workshops and the discussions that took place helped us to gain new insights and to understand the different opinions and arguments within this topic. For example, one citizen pointed out that Pepper's design gave the robot an innocent look, which could potentially make it easy to use to spy on someone (e.g. care workers).

Fig. 7.1 Explaining the functions of the robot. © University of Siegen, Information Systems and New Media

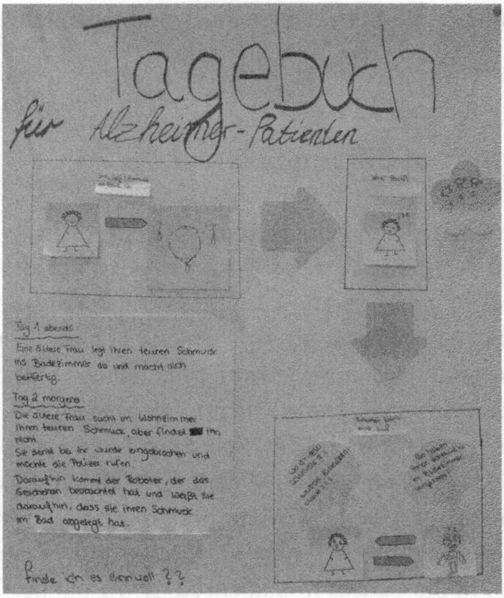

Fig. 7.2 Example of an application area developed by citizen in a design workshop. © University of Siegen, Information Systems and New Media

What was important for us in these events with citizens was that they were based on reciprocity. This means that both sides were expected to take something away with them; citizens should not only be a source of answers to our questions, but should also be given the opportunity to indulge their curiosity. As researchers, we gained knowledge about the application of technology through the exchange, while the citizens gained insights into our research and possible technological solutions for the care sector. Citizens were able to look at the robot, ask questions about it, and receive a detailed presentation of the research results to date. Meanwhile, we were able to gain insights from an important stakeholder group and to further develop our research based on their input.

The results of the citizen participation events helped us to identify a more diverse and general societal picture of care robotics, and to align our research and development with it. A key finding was that robotics should only support the nursing profession rather than replace it (many of the concepts that citizens developed involved three-way interactions between a nurse, a robot, and a person in need of care). This was confirmed in later studies [70, 72]. Robots, we found, should not perform automated processes in nursing and elderly care, but should be part of hybrid teams with nurses or social workers, supporting their work and thus increasing the quality of care.

7.4 Challenges of Citizen Science

In all research, particularly research involving citizens, one should be aware of the need to reach out to a variety of groups. As far as possible, all interest groups and a broad spectrum of opinions should be represented. The variety of perspectives one can get from citizens makes it possible to look at ideas from different angles. Citizen science thus requires a large network of potential contributing citizens.

In addition, researchers should make their work accessible for citizens, in order to create awareness of the research field. Scientific publications often have a complicated structure and use a language all of their own. This style of writing can make understanding difficult for readers who are not familiar with the subject at hand. Scientific research is often prepared only for the scientific community and not for the public, although in most cases it is financed by the public. This has important implications, since it is only reasonable that citizens who both pay for innovation through their taxes and social security payments, and who may directly experience the effects of robotics, should have some say in how robotics develops.

Robotics research has only recently paid attention to the interactive or social potential of robots. This is changing, and we can anticipate that robots will become increasingly present in everyday life, which is why citizens' understanding of social robots, and their attitudes towards them, matter. Thus, a broad network of citizens interested in science would bring societal benefits for all the reasons mentioned above. Making sure that the views collected are representative, however, is not easy. Even so, although this means more effort for researchers, it is important for the future of the research field in the long term, as it improves public understanding of science. One challenge is the fact that citizen involvement in science is often not undertaken systemically within research projects. Public relations often play a rather subordinate role, despite efforts by the various funding agencies, and tend to relate to the publication of results rather than to the participation of citizens in the research process. This problem has already been recognized but not yet solved, despite specific research communication projects such as the "Science Year" in Germany, and Europe-wide platforms created as a basis for citizen science projects [85].

Events involving citizens, we believe, can and should be mutually beneficial. Researchers gain from the insights that a wide range of inputs can bring, and citizens gain a better understanding of what is possible and what might be desirable. Presenting the state of the art in research and development in robotics to the citizen can be challenging, but can nevertheless result in something which benefits everyone. This giving back, if done in the right way, repays the effort of researchers to engage citizens by providing insights which are closer to real life scenarios, thus improving our picture of what the social factors involved in the use of robots might be.

In our experience, it is important to be aware that citizen science projects deliver good and practical results when citizens are involved in the entire research process, not just specific parts of the research. This argument is also fundamental to a concept that has recently become associated with citizen science in Europe, the concept of "Responsible Research and Innovation" (RRI). The core idea here is that societal stakeholders should be involved throughout the research process to incorporate society's values, needs, and expectations into results [329]. Ultimately, public trust in science requires us to move away from the idea of the absent minded professor working away in a lab, wearing a dirty lab coat as a result of too many experiments. A clear way in which we can build this trust, and enhance mutual knowledge, is through citizen science, in whatever form this takes.

7.5 Citizen Science—Further Methods and Resources

There are various ways to conduct citizen science. In the last paragraph we explained our approach, and here we want to present alternative approaches and resources that could be helpful. For example, the website Citizens Create Knowledge [45] was launched to develop such projects. It not only provides comprehensive information about relevant topics, but also helps working groups and networks to start citizen science projects which can be published on the platform. It also enables researchers to participate in other projects, allowing them to gain insights into other work going on in their field which may inform their own work. The website also allows researchers to access an existing network of German-speaking citizens.

While Citizens Create Knowledge is a German platform, the European Citizen Science Association (ECSA) offers a similar service for EU-wide projects (see 7.3). ESCA provides several options for participating in research projects, for example, as a research partner, as a communication partner or by hosting events and workshops. Depending on the required expertise and experience, ECSA can contribute to projects both centrally through ECSA headquarters and through its network of members [20].

Fig. 7.3 The ECSA website

Another method for approaching citizen science is presented by Kirstin Oswald in her article Citizen Science on Instagram. She focuses on the social media platform Instagram, particularly the use of hashtags, to evaluate public attitudes to citizen science [239]. In general, social media can be an important and productive research tool for citizen science. [12] demonstrate this using several case studies in environmental research in their report Citizen Science: Social Media as a Supporting Tool. In doing so, they not only consider social media as a research tool, but also look carefully at the data quality and collection methods in research conducted through social media.

Citizen science requires networking and exchange between science and society [45]. Both social media and websites such as those presented above can be tools for this. They can also help citizen science to become more widely recognized in society and the science world [45]. For successful collaborative research, it is also necessary to integrate citizen science into scientific processes and into educational concepts, as well as integrating its results into decision-making processes [45].

8 Categorizing Social Robots with Respect to Dimensions Relevant to Ethical, Social and Legal Implications

This chapter was published in the i-com Journal, Vol. 19 (Issue 1) as *Tobias Störzinger, Felix Carros, Anne Wierling, Catrin Misselhorn and Rainer Wieching. 2020. Categorizing Social Robots with Respect to Dimensions Relevant to Ethical, Social and Legal Implications. i-com Journal, Vol. 19 (Issue 1), 47–57. DOI:* https://doi.org/10.1515/icom-2020-0005 It corresponds to the P5 in the List of Publications.

Abstract. *The aim of this paper is to suggest a framework for categorizing social robots with respect to four dimensions relevant to an ethical, legal and social evaluation. We argue that by categorizing them thusly, we can circumvent problematic evaluations of social robots that are often based on overly broad and abstract considerations. Instead of questioning, for example, whether social robots are ethically good or bad in general, we instead propose that different configurations of (and combinations thereof) the suggested dimensions entail different paradigmatic challenges with respect to ethical, legal and social issues (ELSI). We therefore encourage practitioners to consider these paradigmatic challenges when designing social robots to find creative design solutions.*

8.1 Introduction

Soon, social robots will increasingly find their way into our households. In contrast to their counterparts—industrial robots—they will not be hidden behind protective fences but will interact and cooperate with their users. Such developments will come with many opportunities. In the near future, social robots might, for example, support the elderly in living independently. However, this also implies that these

F. Carros, *Design, Development and Sensemaking of Human-Robot Interaction in Care Settings*, https://doi.org/10.1007/978-3-658-45233-9_8

technologies will greatly impact our everyday lives. Therefore, they must be designed so as to not cause any physical or mental harm and instead promote our values and contribute to the well-being of individuals and society in general. Thus, it is increasingly important to investigate the ethical, social and legal implications (ELSI) of social robots.

The aim of this paper is to suggest a framework for categorizing social robots with respect to four dimensions relevant to an ethical, legal and social evaluation. We argue that by categorizing them thusly, we can circumvent problematic evaluations of social robots that are often based on overly broad and abstract considerations. Instead of questioning, for example, whether social robots are ethically good or bad in general, we instead propose that different configurations of (and combinations thereof) the suggested dimensions entail different paradigmatic challenges with respect to ethical, legal and social issues. We therefore encourage practitioners to consider these paradigmatic challenges when designing social robots to find creative design solutions.

The paper is divided into the following sections. In the first section, we introduce the general idea of categorizing social robots around dimensions that are relevant in an evaluative context with respect to ethical, legal and social implications. Thereafter, we explain each dimension and explore how different configurations within each dimension can be considered different ELSI relevant "character traits." Finally, we explore paradigmatic challenges that are associated with specific configurations in each dimension with respect to ethical, legal and social issues and aim to encourage practitioners to find creative solutions in their concrete practices of designing social robots.

8.2 Categories to Distinguish Different Types of Social Robots

According to Breazeal, social robots are defined by being "designed to interact with people in a socio-emotional way during interpersonal interaction" [50] (e.g. Fig 8.1).

Various suggestions regarding how to categorize social robots can be found in the literature on human-robot interaction (HCI). The categorization we present below aims at bringing researchers within the ELSI field together with professionals who work in engineering and with designing robots. We believe that our categorization facilitates such an exchange, because it relates ethical, legal and social issues to designable features of the specific robot.

In a recent report, the Office for Technology Assessment, for example, distinguishes three distinct categories: (1) social robots, (2) aid robots for nursing and

(3) robots that help to maintain mobility [169]. Yet, these categories can sometimes be too narrow. The robot Pepper, for example, can be used as a social robot, as well as an aid robot for nursing [243, 315]. The report further differentiates between socially interactive and socially assistive robots. While socially interactive robots have been developed for interactions between humans and robots, socially assistive robots allow communication and serve to promote human-human interactions [103, 194].

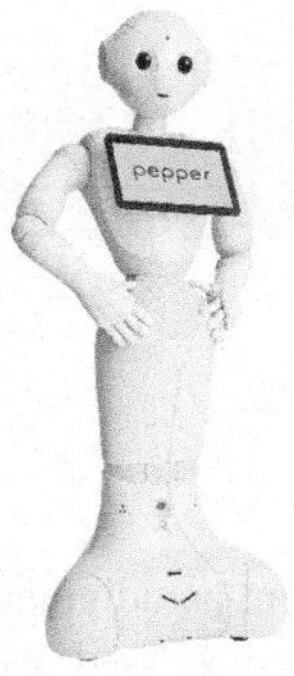		
Pepper, Softbank Robotics https://commons.wikimedia.org/wiki/File:Pepper_the_Robot.jpg	NAO, Softbank Robotics https://commons.wikimedia.org/wiki/File:NAO_Robot_.jpg	Miro http://consequentialrobotics.com/

Fig. 8.1 Examples of social robots

Other categorizations focus on the form of interaction between robot and human [238] or on the specific tasks for which the robot is designed [156]. Our aim is not to reject these types of categorizations but rather to add a framework to categorize different robots to help in evaluating their ethical, legal and social implications. Furthermore, we do not claim that this method of categorization can exhaustively identify every ELSI related challenge regarding social robots but rather that our manner of conceptualizing social robots emphasizes ELSI aspects as part of the robot design process.

To identify ELSI relevant dimensions, we have analyzed sources and materials from eight different research projects on social robotics [57]. By closely examining the outcomes of ELSI workshops and analyzing the ELSI challenges that the projects themselves have identified, we can categorize the different problems and challenges into four different dimensions. The method used here is inspired by the general methodology of grounded theory [297]. We have first characterized the ELSI challenges with which the projects must deal and extracted abstract categories for similar problems. We then conceptualized these categories as gradable dimensions and used this framework to reexamine each project to characterize it according to its specific values in each dimension (Fig. 8.2).

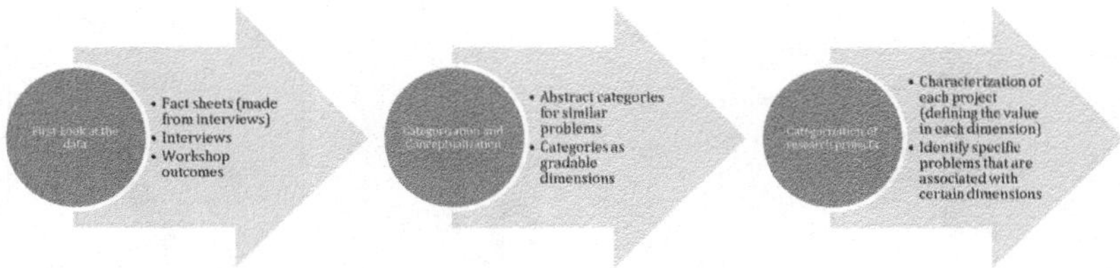

Fig. 8.2 Method used for categorization

In the following, the dimensions are elucidated, and different paradigmatic challenges with respect to ethical, participatory and legal topics are discussed. We do not discuss ELSI challenges specific to the projects upon which our material is based but instead consider general challenges and problems associated with certain design features.

8.3 Categories as Gradable Dimensions

We consider that the following dimensions of social robots liken character traits. This means that each dimension can be attributed an individual degree. Like a person who might be highly extroverted but also possesses further character traits (such as openness, etc.), we conceptualize social robots as possessing different values (or degrees) for different dimensions. This means that a social robot is, for example, not just an emotional robot per se but also has a certain degree of "emotionality" (ranging from 0 to 100). Likewise, social robots are not either wholly autonomous or completely not autonomous but rather have a certain degree of autonomy [211].

In the following sections, we introduce four distinct dimensions (autonomy, emotionality, sociality and impact on competences) to categorize different types of social robots with respect to ELSI relevant features. By considering paradigmatic challenges that can be associated with specific values in each dimension (for example, "high autonomy"), we wish to encourage designers to adopt ELSI considerations as an important part of the design process (Fig. 8.3).

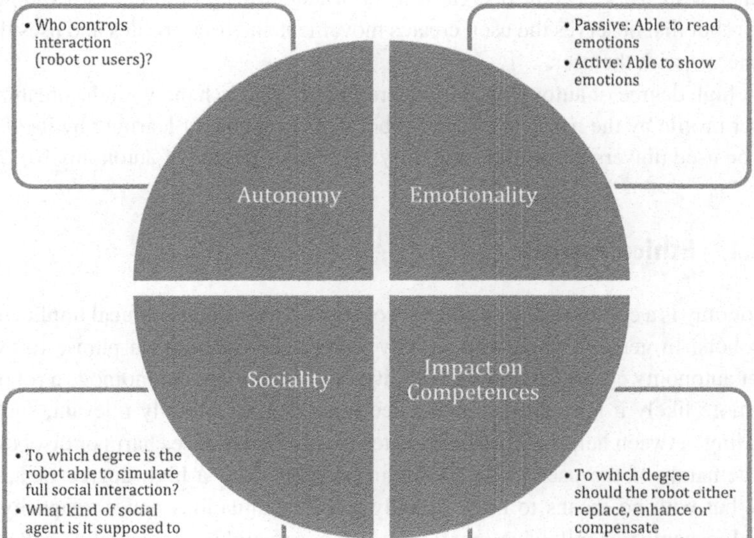

Fig. 8.3 Dimensions relevant to ethical, social and legal implications

8.4 Degree of Autonomy

The degree of autonomy describes who is in control with respect to interactions between robot and user. Whereas a degree of autonomy of zero would represent a robot that is a mere tool and does not have the capacity to act without direct input form the user for each movement, a degree of 100 would represent a robot that makes its own decisions, follows its own plans and can begin the interaction by itself. Autonomy as a gradable dimension was introduced into the discussion of robot ethics by Floridi and Sanders [109] and developed further by Misselhorn[210]. This basal form of autonomy consists of (1) being able to interact with the environment, (2) having a minimal capacity to adapt and (3) being able to change internal states without external stimulus (therefore, this concept of autonomy should not be confused with the much more demanding Kantian meaning of autonomy).

An example of low autonomy would be a typical cleaning robot. Here, the user specifies the days of the week, time frames and locations in which the vacuum cleaner shall work. The next highest level of autonomy would be, for example, a care robot that observes the user, creates movement and user profiles and uses these to interact with the user.

A high degree of autonomy could therefore go hand in hand with the creation of a user profile by the robot, whereas a robot that is capable of learning by itself and can be used universally would exemplify the highest degree of autonomy [6].

8.4.1 Ethical Issues

Autonomy is a central category when it comes to discussing the ethical implications of robots. In general, one could slightly change the well-known phrase to "with great autonomy comes great responsibility," since the more autonomous a robot is, the more likely it is to need to make decisions that are morally relevant, such as deciding between helping a user complete an action that causes harm or disobeying the demands of the user [334]. Designing a robot with a high degree of autonomy but with no means to track morally sensitive situations and no capacity for moral reasoning or following moral rules can create highly problematic challenges [208]. Morally significant situations generally arise much earlier than the debate on autonomous driving or war robots sometimes suggests. In "weaker" scenarios, the robot need not make a life or death decision but rather—for example—decide how to act based on weighing various moral values [15]. And, even if the choices the system must make are not morally prohibited, they still often reflect a preliminary decision regarding ethically relevant values. A technical care system with a lower threshold for alarming relatives may, for example, be more secure but, on the other hand, more annoying, due to possible false positives and, in extreme cases, also lead to a problematic "normalization" of its user's lifestyle [199].

8.4.2 Participatory Issues

A robot's degree of autonomy has an implication on how it should be developed. Autonomy of a robotic system can have a positive effect on the productivity of a robot and can lead to the perception of it as a useful tool. It can also encourage the belief that the robot is smart and possesses an AI and therefore some sort of personality, but at the same time, it can be frightening for users, since its functionality can be difficult to understand [286, 350]. Autonomy of a robotic system has the effect that its users need not take care of the machine at all times; at the same time, this means that they are losing the control over it [289, 299]. This loss of control can result in a reduced understanding of what the robot is doing and a less accurate forecast of what it will do next. The robot becomes less transparent to the user.

This lack of transparency and understanding could lead to diminished trust for the robot and ultimately reduce its effective usage [36, 270]. To avert such a development, it is important to include the users in the development of the robot and tailor its abilities in accordance to the abilities that users expect. Also, limitations of the robot's functionalities should be obvious, to avoid frustration and an extend of trust in the robot that is exaggerated.

Furthermore, it makes a difference whether the robot is used in public or non-public spaces. Public spaces are full of things and situations that are not in the control of the people in it, since they must be shared with others who might have their own agendas and bring things to these places that are not under our control. An autonomous robot might be more accepted in public spaces, since we are all used to being surrounded by situations and things that do not belong to us and because it is possible to retreat from public spaces to a private space or another public space.

The situation becomes a different one in private spaces, such as users' homes. There, it is not possible for them to avoid confrontation. Users must interact with the robot and do not always know what happens with the data that is produced during the interactions, which might me quite sensitive data, like videos or microphone recordings. Therefore, it is crucial that what the robot is doing and planning to do is transparent.

To understand what users, expect from an autonomous robot and be able to design it in a manner that makes it transparent for them, it is crucial to gain a deep understanding of the stakeholders surrounding such systems. For this purpose, we propose the methods of Living Labs [97], a method that makes it possible to observe the usage of robots in real live environments over an extended period [97]. This method allows developers to gain deep insights regarding the usage of such systems and helps them understand what challenges such systems face.

8.4.3 Legal Issues (Privacy)

With a high level of autonomy, it can be presumed that the likelihood of personality infringements arises (e.g. if the robot acts fully autonomously, it makes decisions that are difficult or impossible for the user to predict and therefore lead to a lack of transparency). With self-learning social robots, the traceability of decisions is a possible challenge for practice. The General Data Protection Regulation (GDPR) only provides general regulations (in particular, Art. 22 GDPR) for the use of Big Data and artificial intelligence (AI), which in turn can be substantiated by national laws and the rules of conduct of professional associations (Art. 40 GDPR) [331]. To counteract these transparency problems, initial approaches already exist regarding how AI can be made comprehensible [332].

Concerning autonomous robots making their own decisions in the public sphere, the question is who is responsible for the data processing [209, 315], not only regarding potential liability issues associated with possible damage caused by the robot, but also regarding data protection responsibility. For example, if a care robot goes shopping for the user in a pharmacy, data processing may occur on the way to and in the pharmacy (another question is whether the robot is able/should be empowered to enter into a contract for its user at all). Those affected by data processing (passers-by on their way to, customers in or employees of the pharmacy) must know who to contact to obtain (further) information on the processing of personal data and to assert their rights in this regard. It should be noted that missing or incorrect information (in combination with a provision that depends on the will of the data subject, such as consent or the execution of the contract) could lead to unlawful processing [24, 296]. Who is responsible for data processing depends on who can decide on the means and purposes of the processing. An autonomous robot "decides" for itself, but only within the framework defined for it in advance.

In contrast to use in public spaces, the use of an autonomous robot in a private environment does not fall within the scope of the GDPR, under certain conditions (Art. 2 para. 2 lit. C GDPR—household exemption). One of those conditions is that the manufacturer/seller/programmer of the robot has neither access to the data nor any influence on the purpose and means of data processing. The user himself is responsible on his private property; he decides what, when and how the robot should do/process. If, in addition, no economic or professional connection with data processing exists, the GDPR would not be applicable (e.g. no data subject rights would have to be considered) [9].

8.5 Degree of Emotionality

How we define the degree of emotionality is divided into the two subcategories "active emotionality" and "passive emotionality." A robot with active emotionality can show emotions and trigger empathy. The level of passive emotionality represents the extent to which the robot can read the emotions of its users.

8.5.1 Ethical Issues

A high degree of active emotionality means that technical systems can become objects of our empathy and that we may come to care about them. On the one hand, this need not necessarily be morally problematic, but it also does not automatically lead to an ethically improved technical system, on the other. Whereas interaction and communication with "emotionalized" robots can be more natural and fluent [231], it can also lead to a problematic manipulation of users. This might, for example, be the case with a robot that makes its users do things that they do not want to do by showing negative emotions (crying, whining) until users finally concede. Furthermore, the ability of an artificial system to read our emotional states raises ethical questions. For example, the extensive reading of our emotional state (also based on data that we would not initially associate with it) raises the question of whether we always want to be transparent for these systems.

8.5.2 Participatory Issues

The question of whether a robot should be emotional towards humans (active emotionalization) or should recognize the emotions of the humans surrounding it (passive emotionalization) is highly sensitive and must be answered by its users. In the case of recognizing someone's emotions, sensitive information about the user can be revealed, which the user might not want to be spoken about by the robot or shared with others. In the case of a robot acting emotionally towards a user, some risk exists that it might be manipulative. For example, in the context of care homes and people with dementia, an emotional approach could be perceived as real and unique by the users, whereas in reality, it is a program played for every person who is in contact with the robot, an "industrialization" of personal care. While such a functionality could evoke the positive feeling of being recognized, the question remains whether this is something that is ethically good and desired by the caregivers and the family of the person. A similar situation could arise in any household. If the robot attempts to simulate emotions, a perceived relationship between human and robot could be stimulated, one which is based on program code. Again, this could be perceived positively by the user, yet it should be the user's decision to what extent the robot should be simulating emotions.

To overcome this challenge, it might be useful to let the user engage in a kind of negotiation with the program. Therefore, an end-user development (EUD) program could be helpful [191]. This program would allow users to adjust the robot's emotions, for example, in a way that the robot only shows the emotion of happiness or that it only shows specific emotional reactions to specific users who have authorized such behavior. In the context of passive emotionalization (reading emotions), this program could, for example, define what happens with the gathered data and whether the emotions should be read only in certain time slots, for example, during office hours or breakfast. Furthermore, it could be possible to adjust the reading of emotions to those emotions that other humans can also read but to forbid the reading of things that remain hidden to the eyes of humans, such as a person's pulse or their typing speed. Through an EUD program, users would gain more control of the robot and be able to define the design features of the robot individually.

8.5.3 Legal Issues (Privacy)

One example for a robot with a high level of emotionality could be a robot that helps children with autism to interpret and learn emotions. Emotional manipulation to induce behavior could be an infringement of personality. Regarding the right of personality, it must be considered that, in the field of data protection, a data protection impact assessment must be conducted for such a robot, since "new technology" is used (Art. 35 para. 1 GDPR). This impact assessment identifies possible risks of data processing and tries to minimize the determined risks: The risk exists that the sensitive data of the "transparent patient" may be passed on to third parties (doctors, insurance companies, etc.) without authorization and that the patient may lose control of his health data [94], for example, if a doctor were to send pictures/sound recordings to colleagues for advice. If data were to be unintentionally disclosed, it could lead for example to a loss of confidentiality, health manipulation, physical and mental damage or commercial exploitation. The (health) facility may suffer the loss of reputation or financial damages. To prevent such negative outcomes, adequate technical and organizational measures (TOMs) need to be implemented. Furthermore, certain basic settings that specify maximum data protection (privacy by design/default) should be provided. Examples could be that the release for data transfer to third parties must be actively clicked on in each case [168]. Privacy by design/default could be realized, for example, if the AI generally conducts human-like emotional analyses. Only in case the user activates the advanced function the AI utilizes the full spectrum of what is technically possible to analyze emotions (e.g. analysis of data from facial expressions, posture, spelling or mouse movements).

8.6 Degree of Sociality

Whether and to what degree the robot can simulate social agency is represented by the dimension of "sociality." The paradigmatic background for understanding a very high value of sociality is human-human interaction, which is sometimes defined as the reciprocal interaction of fully autonomous persons. However, other—less reciprocal—forms of social interaction exist, such as the relationships we have with our pets. Social robots with a high value of sociality not only interact with humans but so in a specific manner [281]. They should—for example—simulate being a friend or partner that understands and cares about its users.

8.6.1 Ethical Issues

As mentioned above, human-human interaction is often regarded as the paradigmatic background for creating social human-robot interaction. In cases in which the robot should not merely be a tool or service, the aim is often to design and develop the robot as a companion, friend or roommate. However, research regarding participatory design shows [66, 315] that the robot could be much more effective as a facilitator or mediator for human interaction than as a replacement for social interaction. Keeping this momentary state of art in mind, we should be careful not to create robots with only a very diminished capacity of social interaction and sell them as substitutes for real human-human interaction.

Furthermore, every social interaction is part of creating our individual dispositions and scripts of how we interact with other humans [204, 205]. From this perspective, the question arises regarding what types of effect (diminished) forms of social human-robot interactions have on our dispositions to interact with our fellow humans. Having a friendly robot companion at home that simulates various forms of social interactions but never objects or questions the orders of its users may have negative effects on how users are disposed to interact with others. One must also keep in mind that certain moral restrictions arise from empathizing with robots, even if one knows that they do not really have emotions and only simulate such [30].

8.6.2 Participatory Issues

The role of robots in the social lives of humans is difficult to determine. Some people accept the robot as a social agent, while others repel this concept. Acceptance as a social agent is highly dependent on the predetermined usage and the actual user. The intended role of the robot impacts how it should be developed. If the robot is designed as a functional tool that has specific tasks, it becomes less important to develop the robot applications together with its users, since its task would be to fulfill a specific duty, such as assembling car parts or mowing grass. However, if the robot is designed to have a social role, it becomes crucial to understand the users and their needs and limits.

To do so, it is important to consider the role that users would prefer their robot to occupy. Furthermore, it is important to understand how a robot should act in specific social situations. Determining this can be quite difficult, as some potential users have possibly never been confronted with a social robot. Yet, it is often just these groups that are the preferred targets for user research. Recently, for example, much research has been conducted regarding elderly people in interactions with social robots (e.g. [187] or [256]), as this group presumably needs socio-technical innovation but often have very limited encounters with robotic systems in their everyday lives.

It is crucial to work with users and develop these technical devices together with them to enhance their autonomy [208] and self-respect. To achieve a deep knowledge and understanding of how users interact with the robot, we propose working with design case studies [341, 343]. This method is divided into four phases. (1) The grounded design phase represents the first phase, in which it is important to get to know the users and their living spaces in depth. (2) Within the design phase, a certain use case is designed based on the knowledge gained and developed into prototype. (3) In the evaluation phase, the developed prototype is tested in the envisaged environment. During this time, developers observe the situation, as well as the user's interaction with the robot, and interview the user to reassess whether the intended functionality has been matched. (4) In the iteration phase, based on the knowledge that has been gained in the third phase, the prototype is further developed. Once the adjustments are complete, the prototype is again tested in the envisaged environment and evaluated. This process of iteration is continued until the robot is sufficiently developed to work in a meaningful way, according to its intended social role towards its users.

8.6.3 Legal Issues (Privacy)

An example of a robot with a high level of sociality could be a care robot in a nursing home that tracks the interest and movement patterns of elderly people to determine the best time for each activity with the patient. Possible legal problems in relation to such a robot could be its approval according to the Medical Devices Act, as well as the violation of privacy or human dignity by the surveillance of the patient/user at all times to generate user profiles.

Considering data protection, it must be considered that a specific provision of the GDPR for creating user profiles applies to ensure lawful processing. This provision could be found either in Article 6, for general data, or Article 9, for sensitive data. However, it must be determined what sensitive or, respectively, health data constitutes. According to Recital 35 GDPR, health data "should include all data pertaining to the health status of a data subject which reveal information relating to the past, current or future physical or mental health status of the data subject." Still, the question remains whether the combination of general data, such as the height, weight and sex of a person, is sufficient to represent such a sensitive data point. At least the body mass index could be determined with these parameters, which could therefore be interpreted as health data. It is therefore more appropriate to assume a broad interpretation of the term "health data", since not much (combined) data is required to generate information on the health status of a person ([18, 38]). It can therefore be assumed that sensitive data (health data) must be collected in the context of creating a user profile. The processing of health data is generally prohibited (Art. 9 para. 1 GDPR). If the robot is to process this nevertheless, an exception to this prohibition is required. This could be found, for example, within the framework of Art. 9 para. 2 lit. h GDPR. Accordingly, the prohibition does not apply if the processing is necessary for, for example, preventive health care, medical diagnostics or health care. In addition, processing must be conducted by or under the responsibility of qualified personnel, who must also be subject to professional secrecy (Art. 9 para. 3 GDPR). These conditions must be considering when determining which data should be collected for the user profile.

Alternatively, obtaining consent could also make the processing of health data lawful. To this end, the data subject must explicitly consent to the processing. Within the framework of consent, further requirements must also be fulfilled. For consent to be considered voluntary, for example, no imbalance should exist between the responsible person and the data subject. Therefore, the question of responsibility again arises here. Is the user himself responsible? Could any voluntariness at all exist in a nursing home-patient relationship [180]? It should also be noted that sector-specific regulations take precedence over the general regulations of the GDPR and the Federal Data Protection Act. Some factors are specific to the area of health data—it can be assumed, for example, that consent for the transfer of patient data can only be given voluntarily if this transfer is also permitted by/possible within the law [136]. In addition, the purpose of the processing must be stated so that an informed consent can be given. The question thus arises as to how precisely the purposes for data processing must be described in advance [180]. Health data is regularly used for a variety of purposes (some of them have in advance [180]. Health data is regularly used for a variety of purposes (some of them have not been established in the early stages of Big Data analyses). In addition to treatment, support, care and emergency assistance, purpose could also relate to health management, profitability control, quality assurance, pharmaceutical research or completely extraneous purposes, such as commercial advertising or official actions [331]. It should be noted that all applicable purposes (as well as the legal basis upon which the processing is based, according to Art. 13/14 GDPR) must be communicated to the data subject prior to the actual processing. A subsequent change of purpose is only possible under narrow conditions, which is why it is better to indicate too many possible/conceivable purposes, rather than too few, at the comprehensibly as possible to avoid overburdening the data subject.

8.7 Impact on (Social) Competences

Social robots interact with us, sometimes act for us and sometimes help us interacting with the world. The fourth dimension that turned out to be ELSI relevant involves the question of what impact the interaction has on the competences of users. This impact can be differentiated with respect to the following questions: (1) To what extent is the robot designed to be an enhancement or extension of competences (to enable us to do new things or old things better)? (2) To what extent should it compensate for lost competences (to again do things that one was once able to do)? (3) To what extent should the robot replace competences (to do things that one is still able to do)? The value for each dimension is subject dependent or user dependent, since the "replacement," "enhancement" or "compensation" of competences is directed towards the specific competences of a specific user.

8.7.1 Ethical Issues

Robots can help us to regain competences we have lost, enhance our competences or replace the need to act out our competences. In every case, some caution should exist regarding the dependence on a technical system that is created by introducing and using social robots. In the worst-case scenario, users unlearn important competences when they would otherwise still be able to perform the necessary actions. From an ethical and social perspective, it should be clarified which dependencies are created to ensure that these cannot be exploited.

Another ethically relevant aspect involves identifying competences before attempting to enhance or replace them. One lesson that can be drawn from the discussion on elderly care robots, which should be clear by now, is that the competences of caretakers lie not only in performing certain (mechanical/bodily) actions (bringing food) but also in engaging in social and emotional interactions (see, for example, [113, 244]). Neglecting the latter would entail focusing on only one part of the whole situation, and therefore one might be drawn to the conclusion that replacement without essential loss is possible. Seeing the various competences involved enables developers to set their aim straight: robots might be able to replace or help with heavy and toilsome work to make room for specific human competences.

8.7.2 Participatory Issues

Competences are crucial for every person. When it comes to the development of robotic systems, the questions always exist regarding which tasks the robot should fulfill and whether these tasks are performed solitarily or in cooperation with users [192, 255]. Quite regularly, these developments follow the paradigm of what is possible from a technical point of view, and often, this logic does not consider that people enjoy certain tasks or that they will feel reduced if these tasks are fulfilled by robots and they are thus condemned to do only that which the robot cannot. Therefore, it is important to understand the needs and preferences of the users and gain a deep knowledge on how such robotic systems are implemented in the intended context.

For example, a robotic system that helps users with their grocery shopping, by finding specific items and collecting them from the shelves, could lead to the loss of competence to go grocery shopping without assistance or be able to find products in a supermarket. It must be observed and, in a second step, negotiated whether this competence is something that users would like to keep or can accept the risk of losing.

To do so, it is necessary for stakeholders to participate in product development. Making stakeholders co-creators will drastically deepen understanding about them and be helpful in understanding what competences users would like to maintain and which are less important.

8.8 Conclusion

As we hope to have made clear, we claim not that social robots are good or bad in general but rather that specific configurations lead to specific challenges with respect to ethical, legal and participatory issues. The suggested framework emphasizes that a range of design decisions with respect to autonomy, emotionality, sociality and impact on competences lead to ELSI challenges that must be addressed.

With regard to data protection the provision of correct information to data subjects, methods for explaining AI-decisions, TOMs and the compliance with privacy for design and default requirements are particularly important. Also, all purposes for data-intensive processing must be determined as early as possible and be justified by appropriate permissions.

In respect to participatory aspects, it is important to understand the preferences of the user, as well as the intended usage. To do so, not only is it important to know the user, but he should be integrated into the development, sometimes even being given the opportunity to change the robot following an EUD approach.

Regarding ethics, paradigmatic challenges or questions can be identified for each dimension: (1) High autonomy demands the capacity to make moral decisions. (2) A high degree of emotionality might manipulate users. (3) Which patterns of interaction are reinforced by interacting with a robot that possesses a high degree of sociality? (4) Does the robot promote the unlearning of important competences?

We conclude by emphasizing that practitioners can find creative design solutions for these challenges, as long as they are able to identify them in the first place. It is important that engineers and designers understand that they also make ethically relevant decisions when configurating the robots' characteristics within these dimensions. This awareness should lead to a more reflexive design process, which also involves ethicists. In this way, it is possible to regard the ELSI dimensions not just as constraints but as a motor for better design.

Ethical, Legal & Participatory Concerns in the Development of Human-Robot Interaction—Lessons from Eight Research Projects with Social Robots in Real-world Scenarios

9

This chapter was published in the i-com Journal, Vol. 21 (Issue 2) as *Felix Carros, Tobias Störzinger, Anne Wierling, Adrian Preussner and Peter Tolmie. 2022. Ethical, Legal & Participatory Concerns in the Development of Human-Robot Interaction—Lessons from eight research projects with social robots in real-world scenarios. i-com Journal, Vol. 21 (Issue 2) ,299–309. DOI: https://doi.org/10.1515/icom-2022-0025* It corresponds to the P6 in the List of Publications.

Abstract. *Research on Human-Robot Interaction is increasing as system become widely available and reached a level that enables smooth interactions. Yet, many research projects act in a silo mentality, in regard of participatory, ethical, or legal matters of social robotics. Knowledge about specific challenges is not universal and has often to be transferred from non-robotic contexts. We present findings in the three dimensions: participatory design, ethics for social robots and legal aspects. We accompanied eight research projects on social robots in real-world scenarios for three years. During that time, we spoke, observed, and helped (where possible) the research projects. This gave us specific insights into their work. Our findings work in three dimensions. In participatory design we observed that the trust relationship to users is essential to gain truthful insights and that a mixed method approach is promising. Regarding ethical aspects, we could see that ethical matters should be answered early on. And in the legal dimensions we noticed that the GDPR regulations are a challenge that often requires the help of experts. This work is reflecting on the observation of eight projects and is collecting lessons learned to help future projects and to learn from previous work.*

F. Carros, *Design, Development and Sensemaking of Human-Robot Interaction in Care Settings*, https://doi.org/10.1007/978-3-658-45233-9_9

9.1 Introduction

As the scientific, technological, and social development of social robots has progressed, questions about the ethical, legal and social implications of their deployment have gained attention [66, 210, 295]. Research regarding social robots is now an integral part of the human-computer interaction (HCI) landscape and social robots are increasingly being deployed in real-world settings.

The GINA[1] project worked on the described challenges and formed part of the funded framework "Robots for Assistance", which was supported by the federal ministry of education and research. GINA aimed to oversee and support exchange between eight different research projects, as well as use participatory methods to generalize and deepen the findings related to the design and development of ethical, user-friendly, and robust Human-Robot Inter-action (HRI).

Participatory design [275, 325] has become one of the standard approaches to project-based interaction research and is considered an essential part of uncovering the sociocultural environment of users, so that design concepts properly fit their needs. As its core is the idea that the development of technology should be done together with the people using the technology. However, user-acceptance and satisfaction in no way guarantees that technologies will cause no moral harm or be ethically sound. It has therefore been argued that separate ethical investigation should be undertaken as a central element in technology development [109, 210, 334]. At the same time, participatory design and ethical investigation will count for little if an innovative social robot is not allowed to be used due to a lack of legal conformity, with any breach potentially resulting in hefty fines. This makes GDPR[2] an essential consideration, not only for projects that intend to launch a ready-to-use product on the market, but also for those whose primary focus is research. Accordingly, legal assessment represents a critical component when developing robots within research projects.

Each of the research projects overseen by GINA had its own specific challenges. These may, at first sight, appear unique. However, as will become clear, if one takes a cross-cutting perspective it is possible to identify several commonalities between the projects, which can be applied, in turn, to other research and development projects in this domain. We argue that the findings summarized and generalized in this paper can serve as useful lessons for a wide variety of applications in relation to ethics,

---

[1] GINA is the name of the funded project. In German it stands for: hochwertig Gestaltete INTeraktionsstrategien für service- und Assistenzrobotik (English: High-quality DESIGN INteraction strategies for service and assistance robotics)

[2] General Data Protection Regulation

law and the pursuit of ctogether with a description of the methodological approach each of them adopted. All the projects had a duration of 3 to 3.5 years and have now been completed.

Table 9.1 Overview of Projects with their objectives and approach

Pseudonym	Project Objectives and Approach	Domain
P1	With the help of the robot Pepper, a new form of therapy for the development of socioemotional communication skills in autistic children was developed and tested. The project made use of affective computing to facilitate automatic recognition of emotional markers in a child's voice, face and heart rate to underpin therapy scenarios where interaction between the robot and child could be individually adapted.	Therapy
P2	The use of autonomous robotic systems designed to assist patients in various situations was investigated in rehabilitation clinics. Possible scenarios included transporting luggage or accompanying them to appointments. A stationary robotic arm was also developed that was intended to support staff when handling medical equipment. The overall aim was to build understanding and trust of robotics among patients and staff, so that it can better support all stakeholders in their everyday tasks and activities	Clinic
P3	In this project, a mobile robot was modified into a shopping assistant, so that it could support users searching for different products or in navigating through a supermarket. The robotic shopping assistant also had the capacity to shop autonomously. The overall goal was to implement and investigate different human-robot interactions for different user expectations.	Shopping
P4	To relieve staff in nursing homes and support the residents in their everyday lives, a two-armed robot was developed and studied. In addition to serving and clearing tables, the robot was also intended to be able to offer drinks and snacks, as well as support users in shopping.	Care home

(Continued)

Table 9.1 (Continued)

Pseudonym	Project Objectives and Approach	Domain
P5	The goal of this project was to develop a social robot that could create positive experiences through lifelike interaction, so that the well-being of users could be improved. The robot was conceptualized as a roommate that could recognize the state of mind of a user and react accordingly and empathetically, so that an active dialogue between the robot and human could take place.	Care home
P6	Here, a mobile robotic arm was developed that could safely and intuitively hand over potentially dangerous objects to people with impaired vision, such as filled cups or knives. The robot was studied in a hospital context.	Medical
P7	The goal of this project was to develop generic interaction patterns for resident-robot interaction in the context of elderly care. These interaction patterns sought to be applicable to three different types of robots, so that positive social experiences could be generated according to a user's needs profile through different types of activation.	Care home
P8	In this project, interaction strategies for service robots in private and public spaces were researched. For example, a cleaning robot cleaning the floors in public spaces was designed to interact with passersby as needed and adapt its path accordingly. Possible interactions for a humanoid household robot were also investigated, with the goal being to support users in everyday tasks by serving as a cooperative and adaptive assistant.	Public spaces

From the above table 9.1, it can be seen, that the research projects were highly diverse with regard to their different approaches and objectives. The methodological focus of this paper is upon a qualitative reflexion of these research projects and what lessons they might offer regarding ethics, legal, and participation in the design process. As a part of the exercise, an exchange workshop was organized in which the participating partners were able to share and discuss how their principal findings and experiences related to these concerns. The participants were all experienced research staff employed in academia and actively involved in their respective research projects. Due to the COVID-19 pandemic, the workshop took place online via a collaborative platform.

In addition to the exchange workshop, semi-structured interviews were conducted with stakeholders in the various project partners. As with the workshop, the focus was upon the areas of ethics, participation, and law. In the following, we present and discuss how findings from the research projects speak to three dimensions: ethical, legal, and participatory, each of which is covered in a separate section. The aim of this is to derive learning from the projects, we therefore look at different methods in participatory design and how it influenced the projects, we continue with ethical aspects and how they implemented different concepts like MEESTAR and then look at legal aspects and how GDPR influenced the development of social robots. We conclude by summarizing the lessons learned in a table. Lessons that we learned while observing and discussing with the projects.

9.2 Participatory Aspects

Participatory design covers a wide range of topics and is applicable to many domains and subcategories. Its methodological aspects include user involvement, the relationship between participants and researchers and the sustainable aspects of research undertakings. We summarize below relevant results and lessons learned relating to the application of participatory design in the projects.

9.2.1 User Involvement

Involving users is a core concern in participatory design and was uniformly pursued in all the projects. One particular challenge concerned the appropriate time to involve users in the participatory process. Two strategies emerged in this regard. First, some projects saw advantage in involving users in design decisions early on, as these could then be more easily adapted and improved upon, thereby avoiding fundamental changes later on, incurring enormous effort. However, not all projects saw it this way. So, a second group of projects argued that the distance between imagination and reality meant that early visions of prototypes lacked maturity and that the anticipations of users could lead to them building up unrealistic expectations. This marked out a difference between projects that had developed a social robot from scratch and projects that bought a ready-made system and reconfigured it to their needs, without changing the robot's fundamental design. For the latter, early user involvement was seen as useful for developing use cases, but not for developing more concrete elements. In addition, some argued that the involvement of users gave credibility to the development. Here, it was assumed that if users were involved right from the beginning this would ensure that the development process was user-driven, with the users becoming equal partners in the research and its outcomes. This was felt to increase the likelihood of developed systems being accepted, not only by the users directly involved, but other users as well, because this level of user-involvement could itself be treated as a way of marketing the system's viability. However, this argument needs independent quantitative verification.

9.2.2 Methodological Aspects

Participatory design is characterized by a use of mixed methods, according to the specific objectives to be attained [275]. Mixed methods [179, 307] implies using a mixture of qualitative and quantitative studies. The choice of which approach to adopt first varied across the projects, though almost all used both somewhere along the way. The philosophical orientation of some projects was towards exploring the field quite broadly at first through quantitative research, with the needs for robot development being derived from large quantitative surveys. Others argued that the research field should first be uncovered and understood by qualitative methods. Both approaches are legitimate and just adhere to a different research paradigm. The choice of method primarily depends upon the topic of inquiry [79, 311]. If the focus is on the appearance of a robot, it may be best to first design the robot and then conduct a representative quantitative study to find out, whether the design is acceptable. This was overtly the position of at least one project. Early use of qualitative methods when trying to establish an object of appearance can be problematic because the design is something new and intangible, with the play of possibilities limited by people's imaginations and potentially unclear. When it comes to the design of application scenarios, where the design is focused on the tasks the robot should fulfill, things are different. Many studies focused on robotic systems have demonstrated that the development of applications with users through qualitative approaches can produce positive outcomes (see, for instance, [22, 67, 70, 72, 267]). One of the downsides of qualitative studies is that only a small group is consulted and may only be representative of a small population. On the other hand, one of the downsides of quantitative studies is that respondents may not have understood the questionnaire correctly, so the framing of the questionnaire can be fateful for the results. Recurrent consultation of respondents within short time periods is also very costly. These arguments about the distinctions between qualitative versus quantitative research are, of course, not new, but they were replayed in the observed projects and had an impact upon the results obtained.

Another qualitative approach adopted by two of the projects was Wizard-of-Oz techniques [87, 95]. The original goal, here, was to overcome technical limitations by making it appear to a user that they were dealing with an autonomous robotic system, while in the background a human being was con-trolling the system and what the robot said or did. The argument put forward, in this case, was that this method allowed the researchers to obtain a robot's perspective and thereby grasp both its limitations and its possibilities. This particularly helped when implementing adaptations to a robot, by making their impact easier to understand, as the person controlling the robot was seeing immediate changes in the reaction and interaction of users. It also allowed the projects to better understand the role of the robot and adapt it accordingly.

9.2.3 Trust Relationships

Trust is an essential part of participatory design because the researchers depend on the honesty of the users. This makes the work of establishing a trustful relationship critical. The projects stated that earning the trust of the participants was one of the greatest challenges. Paradoxically, the results in projects with a qualitative focus were sometimes overly positive because the subjects did not dare to voice criticism out of politeness. This limited the scope for iterative improvements in the design.

The relationship that evolved between the participants and the researchers over the course of these projects made it more likely the participants would express positive opinions. They didn't want to risk offending the re-searchers by criticizing their work, so they sought to tell them what they thought they wanted to hear, rather than what they really thought of the design. This was not always the case but could be observed on different occasions when several parties were interviewed about the same interaction between a robot and a human. Care residents tended to state positive views of the interaction during interviews, while care workers put forward different views about what the residents liked and disliked. This was especially the case when the researchers were not choosing test subjects themselves and at random, but rather obliged to let host institutions, such as care homes, select the participants. This tended to drive participants to only reflect on the positive aspects of their experiences, while the negative ones were set aside. In qualitative research, the building of trust between researchers and their subjects is important. This ensures that the subjects are able to become accomplices in the development of robotic systems and that they are willing to accept this role.

A full and honest relationship provides a foundation for making it clear that criticism is desirable, without social consequences, and that it will assist in improving the design. At the same time, however, it must be considered that the role of such accomplices should not be overstretched by asking too much of them. It is therefore important that, when doing things like conducting interviews with older users, they should be kept as short as possible, with straightforward expressions and, if possible, not only online, but also in person and in places that are familiar to them. Online formats have proven to be useful, but they can be difficult to handle for users who do not have much affinity with technology. It is also harder to build a trustful relation online, leading to skewed results (as explained above).

In studies conducted with vulnerable groups, such as older people, variability in participation also must be accepted as par for the course. In the care settings, for instance, residents and care staff may temporarily or entirely drop out of a study for health or private reasons, in which case new arrangements will need to be made and fresh work devoted to create trustful relations to the new participants. In qualitative research this can result in new user groups putting forward completely different ideas and requirements. As qualitative research always works with small groups, it is not given that a newly recruited group will come to the same ideas and reactions towards the social robot.

9.2.4 Sustainable Development

The sustainability of projects relates to the continuation of deployed developments and to how functioning developments are left after the initial research phase (in this context, the term 'developments' refers to research objects, so here it is about robots and the software developed for them during the research project). One of the problems with research in real-world environments is that users become accustomed to developments and integrate them into their everyday practices. When the research project comes to an end, the developments often collapse because the necessary personnel and resources are no longer available. This is especially the case when a large part of the research is carried out by universities, where many research projects are only funded for a limited period of time. For an participatory design approach this is not always a good result, since the social robot was developed together and participants have been accustomed to the robot. It also stands in the way of future work, as participants might take it negatively that everything ended as soon as the funding of the research project finished.

Within the projects we were observing, we noted two principal approaches to sustainability. For some, the project consortia had been formed in such a way as to ensure that the development was taken over by a commercial company at the end of the research project. Here, the consortia consisted of several partners with different backgrounds, some from research institutions, others from companies. The role of the company was usually, among other things, to further develop and market the product after the research phase was completed. This approach provides a certain sustainability for users because, once a product is marketed, they will continue to have access to it on the open market. The second approach focused on publishing the results, so that external researchers and developers could benefit from the findings and take the work forward. For this, apart from published papers, the documentation of open-source approaches and the creation of long-term repositories can help to underpin sustainability. Public relations via social media and individual websites can also play a role here. We found that both tended to be used in tandem. Publishing is usually a requirement for research projects; bringing the product to the market is in the interest of the companies that are involved.

All projects evinced an interest in sustainability. Especially once they had worked intensively with end-users. However, not all were equally able to maintain it because it often depended on business decisions outside of their control. The problem for the research projects was that the developments need to become regular products in order to ensure that it becomes permanently available to end-users. Offering products is mostly not possible for research institutes, companies have to offer these products. Companies were included in the research projects, but not all of them decided to create products out of the research.

9.2.5 Lessons Learned and Possible Measures

Taken together, our results suggest a few lessons that can be instructive for future activities in the field of participatory design for Human-Robot Interaction. The lessons are based upon the results but are interpreted more broadly to ensure their generalizability. They indicate that the development of social robots requires a holistic approach with a strong focus upon end-user participation. The lessons are presented in the following Table 9.2.

Table 9.2 Lessons Learned on Participatory Design

Leasons Learned	Description	Possible Measures
Qualitative user research is suitable for application development.	Qualitative studies provide the appropriate level of detail to understand the complex interaction between humans and robots. By building on these, robot development can be undertaken in a sensitive and nuanced way.	Incorporate qualitative approaches throughout the research process.
Use Wizard-of-Oz approaches to understand a robot's potential role.	Wizard-of-Oz approaches help researchers grasp a robot's perspective and the role it might play. By controlling a robot, they can acquire a deeper understanding of how it interacts with people.	See Wizard-of-Oz approaches not only as a simulation method, but also as a resource for understanding Human-Robot Interaction.
Develop trust relation-ships with participants	In participatory design, study participants need to be as open as possible with researchers, so that the feedback is frank and honest. This makes improvements based on the feedback more robust. As research regarding social robots often takes place in relatively sensitive domains, developing trust relationships can be especially important.	Fully communicate planned activities early on and verify how stakeholders will be involved, so that replacements can be found if any of them are likely to be absent.
Think about sustainability from the outset.	The sustainable integration of a developed product can be difficult for universities to implement. After a project has run its course, participants often no longer have access to developed products because they are no longer available on the market.	Before a project begins, clarify how its outcomes will be maintained and who holds rights to future development.

9.3 Ethical Aspects

Ethical concerns range from anticipating and preventing moral harm (for example violations of the autonomy of users) to ensuring that an envisioned robot will genuinely contribute to the lives of users. Different projects tackle these issues in different ways. Some involve consortium partners who are specialized in ethics/moral philosophy and who are able to review the ethical aspects of prospective developments independently before they are implemented.

In the projects we observed, most of them tended to have recurrent ethics workshops, but without there being a consortium partner with an ethical background and who deals exclusively with ethical issues in the project. When conducting such workshops, it is highly common to use the MEESTAR workshop model. [200]. We also found that some projects drew upon external experts for this task, while others made use of internal resources. Apart from using MEESTAR as a framework for dealing with ethical questions, most projects also undertook the following activities as part of their 'ethics toolbox': expert interviews; the creation of user journeys; and structured exchange between project partners and potential users and stakeholders. The latter could involve (online) surveys, qualitative interviews, living labs and/or user testing.

9.3.1 Ethical Methods Used

When asked (in the exchange workshop as well as in the semi-structured interviews) about the extent to which the methods the projects used to tackle ethical aspects were helpful or profitable and what lessons could be drawn from the experience, the following picture emerged:

Successful ethical analysis was strongly associated with the degree to which potential users and stakeholders were involved early on. Most projects felt it was essential to involve as many partners as possible from the outset, so that they could cultivate feedback and, where necessary, iteratively adapt their envisaged solutions.

A vital factor influencing the successful handling of ethical aspects was seen to be the timing and scheduling of relevant activities. The projects felt that ethical aspects as with all ethical, social, and legal implications (ELSI) should be addressed as early as possible because technological or functional changes became harder and more expensive to tackle later on. It was even suggested that an ethical evaluation, together with assessment of a project's overall ethical orientation, should be undertaken before project start. This should involve answering questions like: "What is the overall goal of this technical solution or product we are about to develop?" or "How and why is this solution beneficial for the users and society in general?".

The projects also highlighted the importance of assigning a dedicated ethics work package in the project plan, with all of the relevant procedures, timelines and responsibilities properly identified. It was therefore suggested that ethics should remain an essential part of the application process for the funding of technological development projects. It was also argued that incorporating ethics experts in a project team should be standard practice. There was a general agreement that more time, attention, and responsibility should be assigned to the handling of ethical matters. Apart from this, it was pointed out that ethics itself lacks established and agreed-upon methods that might be systematically applied and that there was an urgent need to develop standardized ethics procedures.

9.3.2 Lessons Learned and Possible Measures

Before going on to list the overall lessons learned relating to ethics table 9.3, it is worth examining more closely two of the points that arose during our analysis. One concerns the understanding of ethics partly expressed by the projects, the other concerns the suggested need for standardized methods. In our opinion, both lead to the same conclusion.

Table 9.3 Lessons Learned on Ethical Aspects

Leasons Learned	Description	Possible Measures
Engage in consideration of ethics early on.	Addressing ethical issues at an early stage (preferably when applying for funding) ensures that potential moral harm can be averted and that the project can make a meaningful contribution to people's quality of life.	Undertake ethics workshops at the application stage, asking questions about the meaning and purpose of proposed developments, well as their moral implications.
Treat ethics as a separate work package.	Dealing with ethical aspects needs time and resources. Not only should each project partner have work packages dedicated to this in their work plan, but one partner should also take lead responsibility for their realization.	Establish clear responsibilities and work packages for ethics.

(Continued)

Table 9.3 (Continued)

Leasons Learned	Description	Possible Measures
Involve user groups and stakeholders as soon as possible	Ethical evaluation also needs to take into account how users feel about a proposed product. As user consultation may necessitate changes of direction, their early involvement is to be strongly advised.	Apply participatory methods at an early stage.
Ethics is not just about well-being.	Ethics goes beyond focusing on the well-being of potential users. It must also involve making justifiable and reasoned decisions that are based on justified principles.	Use ethics experts to facilitate joint reflection and accept their guidance.

First, some of the projects saw engaging with users as a way of addressing the need for ethical input. Even if it is problematic *prima facie* for a technical solution to be directed against the well-being of users, a mere survey of user satisfaction cannot replace ethical reflection. Ethics is based on general principles that require continual self-reflection and decisions cannot be legitimized solely based on user satisfaction. So, in addition to ensuring the well-being of users, we also advocate more general reflection upon the ethical values, moral principles and social acceptability attaching to specific activities and proposed solutions.

Secondly, when it comes to the perceived need for a standardized method, although this is understandable, as is the associated desire for ethical guidance, it is in tension with notion of ethics itself. For a start, ethical challenges and opportunities vary greatly depending on the proposed technical developments, making it essential that they be evaluated individually. In addition, ethical evaluation is about making reasoned and justifiable decisions. Reasoned and justifiable decisions cannot be made by working through a checklist or a standardized procedure; they require argumentative judgment and qualitative analysis tailored to each specific case. The main issue concerns the application of general ethical principles and values to the specific situation and the making of reasoned trade-offs.

This underscores the need for cooperation with dedicated ethics partners, such as trained ethicists or philosophers. Such parties can deal with ethical aspects and be relied upon to use their expertise and methodological knowledge to make ethical effective considerations in consultation with the project team. There are important ethical aspects that cannot be identified simply by using user acceptance testing and basic assessments of ethical acceptability. Trained ethicists know best how to facilitate reflection on project objectives and the overall justifiability of envisioned products, thus adding another critical layer of evaluation.

9.4 Legal Aspects

With the introduction of the new GDPR, rules have been laid down on how, for example, data relating to test participants must be handled, so that they are adequately protected from third parties. This has resulted in considerable additional effort for researchers and represents a particular challenge for the design of participatory studies. Regarding the participant documents, data protection is seen as a very complex issue that goes beyond any current research interests. External advice from data protection experts is therefore required and this can involve significant extra expense.

9.4.1 Early Familiarization with Data Protection

For the design of technology and its use to be compliant research projects must adhere to GDPR regulations, as soon as sensitive user data (e.g., health data) is evaluated.

Some of the projects had a legal expert that they could contact, with questions regarding GDPR in either their consortium or in their associated companies or universities. They were able to contact these people on a regular basis and therefore, had no problems with consent, complying with privacy by design requirements and the use of robot speech recognition, or creating the inventory of their processing activities. Projects that were able to use existing templates or draw upon experience from previous projects also had less trouble with consent, establishing records of processing activities, or in establishing technical and organizational measures.

However, there were also projects that did not have a direct contact person or did not realize until too late that they had such a contact person. In these projects, the designated persons had to deal with the topic for the first time. If the person responsible for data privacy must first familiarize himself or herself with the subject area, he or she may not be able to correctly estimate the scope of the work to be performed in the context of data privacy. For instance: One project did not consider who else might be interested in the data when obtaining consent, which made it impossible for collaborating partners to use the data collected. A new data collection with newly formulated declarations of consent had to be carried out, which in turn led to a loss of time. Another project had problems with consent for a study. The consent only covered audio but not video recordings, which meant that the videos could not be used to evaluate the study, as it is not possible to obtain consent retrospectively. If a corresponding study, then must be conducted again, significant time problems arise. Time problems also arise, if, for example, processing times are not considered in the data protection concept (one project had found that, amongst other things, the pixilation of videos could take an unexpectedly long time).

The correctly estimate of the work and the fact that potential difficulties in the context of data protection are not recognized as such can lead to problems: For instance, one project indicated that person recognition might have been implemented via RFID chips instead of biometric recognition if it had been known earlier that the GDPR compliant identification of persons could be a problem. A second project indicated that a learning system or database for voice control could have been realized through real-time processing if it had been known early on that there could be GDPR complications with the voice control that was initially planned. Also, people who are not familiar with the subject sometimes do not take important details into account or simply overlook them. For in-stance: Amongst the projects we were observing, not all of them were aware that a register of their processing activities had to be created in accordance with GDPR Article 30. It was also sometimes necessary to clarify when consent needed to be obtained and at what time certain kinds of data must be deleted. Another project was unclear about what personal data actually is, when information regarding Articles 13 and 14 of the GDPR must be provided, and what this information actually is.

If the person responsible still has to familiarize himself with the topic of data protection, it is possible that problems are not recognized at all or are recognized too late. If they are identified too late, there may not be time to find or implement alternatives. Further, legal questions within the consortium might not be answered competently and, more importantly, in a timely manner, so that delays can occur more frequently.

Overall, our observations of the various projects suggested that GDPR concepts should be directly addressed at the start of a project and then regularly adapted throughout its course. All too often, it is still assumed that a simple declaration formulated at the end of a project is sufficient to meet the requirements of GDPR. A rude awakening then comes when technicians inform the consortium that the corresponding requirements (for example deletion regulations) can only be implemented with an enormous expenditure of time and that, had these requirements been considered earlier, they could have been implemented much more easily. Therefore, a responsible person who is familiar with GDPR requirements should be on board right from the start. If such a person must first become familiar with the subject area, this has a negative effect on the timeline involved in moving from concept to design.

9.4.2 Handling GDPR-based insecurities

The projects dealt with legal ambiguities in a large variety of different ways. One project had an expert in the consortium who recommended avoiding anything critical. This is also the approach that most projects (six out of eight) have taken once they have faced difficulties: In one project, the initial goal was to distinguish groups of people by using cameras. The idea was that, when the robot approached people, it would be able to differentiate whether they were children, handicapped people, elderly, etc. This would then guide the distance the robot maintained from them. However, there was a concern that this kind of differentiation might not be GDPR-compliant (i.e., amounting to the collection of biometric data). It was therefore decided that this functionality should not be implemented. Another project decided that patient identification could not be implemented because of the principle of data minimization outlined in GDPR Article 5. Also, the recognition of visitors is a robot function that was not implemented by one project due to GDPR requirements. Personalized voice output, such as "Hello Dave, how was your surgery yesterday?", was also not implemented because of the sensitive data required for this. An initial plan to capture camera recordings in a patient's room was mentioned as another robot function that was set aside be-cause of the GDPR. Another project stated that a permanently active camera was not implemented. A learning system and an internet connection for the robot were also not implemented due to GDPR requirements by another project. One project even subcontracted the legal considerations to a lawyer, who advised them to have the courage to try something new, as court proceedings would probably be unavoidable anyway when working in such an innovative domain.

Another way to deal with GDPR-based insecurities is to look for alternative functions of the robot: After a project decided against the implementation of biometric recognition of people due to GDPR requirements reassessment of the use case made it clear that it would be sufficient to recognize people by their clothing instead of biometric data. Another project stated that the robot, which is also supposed to go shopping for users, cannot bring the ordered goods to the ordering person as initially planned due to the GDPR requirements. As an alternative, the robot now brings the goods to the place where the order was placed. Also, due to GDPR requirements, the order is not saved as an audio file, but only as a text file, made possible by the use of speech-to-text.

With the benefit of having had an overview of all the projects, we would recommend proceeding in a way that will not cost as much as a lawsuit, while still avoiding having to discard ideas. Our suggestion is to seek help from public bodies such as supervisory authorities at an early stage. If the supervisory authority gives the green light, the project can go ahead without changes. If not, there is still a chance that the supervisory authority will give guidance regarding what adjustments are necessary for an implementation to proceed. If the response from the supervisory authority is wholly negative or not likely to be received soon enough, other approaches (such as find-ing alternative functions for the robot) can still be considered.

9.4.3 Lessons Learned and Possible Measures

In the following table 9.4, we present some overall lessons learned from across the eight projects that may be helpful for future projects when dealing with GDPR considerations in the social robotics domain. We also give some possible measures that might be undertaken to address the issues raised.

Table 9.4 Lessons Learned on Legal Aspects

Leasons Learned	Description	Possible Measures
Think about GDPR from the start	Early processing of potential GDPR requirements (preferably in the application phase) helps to ensure that decisions do not have to be reversed and that targeted development can be undertaken from the start.	Involve legal partners in the application phase and requirements analysis.
Allow time for data protection requirements to be properly met.	Concepts associated with data protection can be time-consuming and require expertise to be properly realized. If this is not taken into account at the outset, additional expenses may be incurred.	Allow dedicated person-months to cover the handling of data protection concepts
Data protection concepts should clarify whether the data will be used for further research or cooperation partners so that appropriate precautions can be taken	If a partner is interested in using collected data later on, obtaining further consent can be complex and is not always possible, necessitating a new round of consent and data collection. It is important to know from the beginning who might be interested and to build this into the process.	Plan for potential extended data use and canvass partners for any prospective interest early on.

9.5 Conclusion

This paper has presented an investigation of the participatory design, ethical and legal aspects of eight different social robot development projects. We have explicated the lessons learned for these categories regarding particularly significant issues that arose and have looked at what measures might be undertaken to handle them. These lessons can also serve as guidelines for future projects in the context of robot development. The paper has uncovered the following take home messages for robot development projects:

With regard to participatory design, a common feature of the projects was a desire for sustainability, such that the work undertaken could be continued for those who participated in the development. This also feeds into the relationships that need to be built with study subjects, so that a basis of trust can be established. In some cases, the same participants have worked for years to facilitate robotics research, so, having a development stop before the product is ready is unsatisfactory for both users and researchers. At present, innovations may not be available to participants even when research goals have been achieved.

From an ethical perspective, consideration of the ethical implications of a development needs to start as early as possible. This is best encompassed within a clearly defined work package with assigned responsibilities. It also helps significantly if there is a competent ethics partner in the consortium. It also needs to be remembered that, while participatory approaches might be an essential part of ethical evaluation, they cannot be a substitute for broader ethical reflection.

In relation to legal matters, early consideration of GDPR requirements is essential, including the identification of a suitable contact person for questions regarding GDPR. GDPR concepts relevant to a project need to be considered from the outset. This includes assessing whether cooperating partners might have an additional interest in collected data, so that this can feature as a part of the consent process. If doubts remain regarding the interpretation of individual GDPR requirements, the first step should be to try to obtain feedback from supervisory authorities, rather than risking long legal disputes or giving up on potentially valuable planned robot functionality.

In summary, rigorous research in real world environments underlies difficulties. Participating parties can change (e.g., care homes), new hurdles can emerge (e.g., GDPR), and framing conditions can be subject to upheaval (e.g., COVID-19). This makes it important that research projects retain some flexibility. However, potential disruptions cannot be easily anticipated in project proposals and incorporated into work packages. It may therefore be necessary to build in buffer periods that allow for setbacks or unexpected challenges.

One final consideration to raise is that many challenges are cross-project challenges, rather than issues that concern just one single project. Research in Human-Robot Interaction underlies specific challenges, as robots are less flexible than software and users often have specific assumptions of the abilities of robots, in addition, as it is an emerging technology ethical and legal aspects are not common knowledge, many problems have been addressed already but not openly enough be discussed within the research community. Adopting a silo mentality between competing project consortia is therefore not always wise. In this respect, a greater level of open science and sharing between projects is clearly desirable, so that the same problems do not have to be solved again and again by different parties.

Part III
Making Sense

The third section of this dissertation elaborates on the results of the previous chapters and draws conclusions and recommendations for the HCI community. Chapter 10 discusses the results and contrasts them with other research findings, while also offering personal reflections on the topic. Chapter 11 outlines the limitations of this dissertation, which are mainly due to the qualitative practicebased research. Chapter 12 concludes the presented work and offers my perspective on the field as well as thoughts on future research.

Discussion 10

10.1 From Social Robots to Practice Research

Robotic systems have made tremendous advances in recent decades. As previously described in 2.1.1, we have seen a rapid development since 1950. Robots have developed from simple machines that could manipulate simple things by grasping and placing them, as Zamolla et al. describe it [349], to machines that try to become social by detecting other movements in their vision and reacting to human shapes. The social robots of today have smart software systems that are partly able to understand what humans are doing and saying and that are partly able to react to it. With these new capabilities we are seeing an increase in research on how to use them in socially oriented fields of work. One of these is the care setting, as this dissertation shows, but there are also other fields, such as restaurants and food service [246, 283], libraries [227, 347], hotels [220, 242], and many more. The differentiation can also be made about user group. For example, beside the user group of older adults there often also is research about children (e.g. [60, 164]).

The challenges facing robots in practice is, to some degree, similar to earlier challenges in the CSCW field. It would be too narrow a view to look only at the interaction between one robot and one individual. Rather, it is important to understand the context and how relations are interwoven; therefore, we must take into consideration a group of persons and institutions and look into community of practices (COP) [159, 322, 326]. We further need to take into account that interactions

---

Supplementary Information The online version contains supplementary material available at https://doi.org/10.1007/978-3-658-45233-9_10.

F. Carros, *Design, Development and Sensemaking of Human-Robot Interaction in Care Settings*, https://doi.org/10.1007/978-3-658-45233-9_10

with a robotic system need to be managed and moderated, which means that timetables and schedules must be organized. Also, the skills on how the robot must be shared within the different groups and, lastly, different work roles must be coordinated. Only with this information can we understand how software for robots should be designed.

In the following subsections I will expand on these challenges and show, from an socio-informatics perspective, the conclusions we have reached, but also the questions that remain open.

10.2 Perception of Social Robots in Care Settings

The design of applications with participatory design methods for social robots is a challenge in itself. Perceptions of what robots can do have, to some degree, been shaped by fiction [271]. Robots are a common theme in popular fiction like movies and novels and have influenced common beliefs of what a robot is able to do [141]. Robots in fiction are often displayed as intelligent and strong, and they often have superhuman abilities. Paired with the humanoid shape of social robots like Pepper, such depictions lead to a widespread belief that these machines are intelligent beings with a wide range of physical abilities (e.g., the Pepper robot has fingers, which this leads to the belief that they work the same way as, or even better than, human fingers). These beliefs about social robots made the process of participatory design difficult for us, as the capabilities and limitations of robots must be carefully explained before it is possible to work in a participatory design process [70]. So, when working with groups or individuals to develop applications for the robot (e.g., cognitive activation games) we first must carefully explain the abilities and limitations of the robot. This process is not unique to participatory design and can also be observed when working on applications that run on smartphones or notebooks [93, 161, 335], but the challenge with social robotics is particularly salient because people's ideas of what a robot is able to do are heavily shaped by fiction. In one paper, we collected some of these fears and could show that they are manifold but mainly connected to the loss of jobs and of being monitored [66].

Another challenge to the development of social robots are the limitations of the system itself. Just because a robotic system has a specific shape (e.g., fingers) it does not mean that they are functional. Specific shapes that indicate functionality are often developed only to create a better interaction. We can observe a mismatch between the design of the robot and its actual functions. According to the "uncanny valley" theory of Mori [215], users are more likely to interact and accept a robotic system that looks human-like. But even if anthropomorphizing a robot creates a higher acceptance, there is also a danger in it, as users tend to overestimate the capabilities and sometimes trust its abilities too much. To cite one example: a social robot was developed to patrol in a parking lot. The robot was designed to create a feeling of safety, had "POLICE" written on it, and featured an emergency button. At one point, a real emergency developed and one person tried to inform the authorities by pushing the button, only then to realize that the button was not working and that the robot was not connected to the local authorities. The robot was of no help and did not react accordingly, instead it tried to continue with its programmed routine and continued the programmed path, while ignoring the emergency at hand and telling people that they need to keep the area clean. The person ended up calling the authorities with a smartphone, with longer-term consequences for trust in social robots [290]. The person had trusted the robotic system too much, as the design of the robot was promising more then it was capable. This was also happening in research projects about HRI; it is called "overtrust" and can sometimes be observed even after participants have seen a robot fail before [80, 260]. Something similar could be observed in our work in the care home, although not with the same severity. The residents repeatedly tried to talk to the robot, expecting it to understand them. They kept on doing this even after working with it for months. We told them several times that the robot can only react to specific questions like "how are you?" But because they experienced these interactions ("how are you"), they assumed that more was possible. The robot was not able to understand what the residents were telling it, but because it would sometimes react, they assumed it could always understand them and vastly overestimated the system's abilities and intelligence. This is something we often see in the development of robotics and AI: people see one single action of the systems, like playing GO [76, 119] and assume that because it can perform this specific (narrowly defined and rule-based) task, it can also perform other tasks. The classic FITTS HABA MABA comparison (see [47]) distinguishes the elements of tasks that make human beings suited to some tasks, and machines to others. Ordinary people, however, lack awareness of these distinctions. The point is that the differences underpin the so-called "allocation of function" principle: without a clear understanding of the limitations of both parties, functions cannot be efficiently allocated.

Overall, for the perception of social robots, three challenges can be observed: (1) the image of robots is shaped by fiction, (2) the design and the functionalities of a robot often do not match, and (3) users tend to think that robots can do more than they actually can and therefore put too much trust in them.

10.3 Attitudes toward Robotic Systems in Care Settings

In the context of this dissertation, four large stakeholder groups have been the focus. First, of course, are the residents of the care homes. This group was the most important to us and all early development was centered around their needs and preferences. Second was the group of people working in the care homes. This is a diverse group of different professions that includes social service workers, caregivers, assistant caregivers and the management of the care homes. We started to call them "care workers" because the professions are diverse but all work with the same residents in one facility [72]. Third are the secondary stakeholders, people and organizations that are affected or influenced by the usage of technology like robots in the care home but do not hold immediate stakes in the care homes [82, 183]. Examples include insurance companies, NGOs, the municipality, IT companies that produce software for the care sector, architects, and doctors. The fourth and final group are citizens, who were already discussed in Chap. 7 and 3.3.

The first group are residents of care homes. Results from this group are difficult to generalize as the group is very heterogeneous. Most of the residents are older adults, but we have also worked with younger residents, so the category of age does not narrow it down. If we look at the largest group, older adults, we must acknowledge that there is high diversity within this group. Psychologically and physically, the groups differ significantly [150]: although it may be possible to do sport exercises or do certain cognitive activities with one resident, that does not mean these will be possible with the other residents as well. At the same time, the perception of what is seen as an "old person" differs within this group as well; for some it is age, while for others it is health status [106, 174]. A better category might be the word "frail" [336], given that all residents who worked with us had certain health issues and relied on the services of the care homes.

With the participating residents, we saw that they were generally open to using such a technology as the social robot Pepper. Once showed how to use the robot, the residents were willing to use it and participated in the interaction. They were singing with the robots, doing physical exercises together, or engaging in cognitive games [70, 72]. At least, this was true of those who worked with us (voluntarily, with no compensation). But we also saw that some had reservations and needed time to accommodate to the technology. For example, we had one resident (Mrs. S) who preferred not to participate in our studies. One time we used the robot in a space where most of the residents were eating (restaurant); in this common area, the robot interacted with other residents, and while doing so Mrs. S could observe the robot and its behavior from far away. This made her curious and after some time she also asked to interact with the robot. This resonates with our other results showing that interaction with the robot tended to increase with time and that the residents were learning from one another, a well-researched effect in other fields of technology usage [226]. Often the robot would interact with groups together with a moderating person. In these sessions, cognitive and physical exercises would be done that were programmed as applications of the robot. Organizing regular sessions helped, as the residents learned what was expected of them and it became a routine.

Fig. 10.1 Examples of robot Pepper. (Source: https://commons.wikimedia.org/wiki/File:Pepper_the_Robot.jpg)

We could regularly observe emotional responses to the robot. Some residents tried to hug the robot and winked at it. Such emotional responses were predominantly observable among the female residents, whereas the male residents were more emotionally reserved and instead asked us many questions about how the robot worked and what machines it used. This emotional response was most likely triggered by the design of the humanoid robot, Pepper (see Fig. 10.1), that to some degree is designed with childlike attributes, with its large eyes and short stature of

1.2 meters [243]. However, it could also be attributable to the higher ratio of women to men in the care home.

During this work, it was noticeable that the attitudes from people working in the care setting (Group 2) toward robotic systems in care were polarized. Some think it is the solution to the increased pressure on the care system, while others believe that it is unethical and a development that should never happen. This polarized opinion usually changed once the robot was presented and the people were involved in the participatory approach of developing the software of the robotic system. Those with critical attitudes usually saw that the robot could only assist with the work and not take over completely. But the people who thought that the robots could be a solution to many problems also realized that, although robots can be of help, they can only offer some assistance and are not an absolute solution to the labor crisis in the care system.

Another notable observation was that social interaction is seen as a desirable aspect of the job. Many people entered the profession to be close to other people. Thus, it was unclear in the beginning how a robotic system built for social interaction would be seen. However, the care workers still saw it as something useful because social interaction is more than just the applications of the robot; it happens all the time and sometimes in very short sequences, like riding in the elevator together. So, although the robot created new social interaction, it did not take away this part of the work.

In the years of working together, we saw that care workers use a complex set of social skills and practices to fulfill their job duties while, at the same time, they need to be aware of a handful of things at once because they often work with several residents and must perform several tasks in parallel. They also often have highly developed empathic skills and often understand the needs of the residents in the situation before talking to them. But part of this empathic behavior is also the ability to be able to distract residents (e.g., in the case of residents with dementia, it sometimes happens that they want to go home or miss people who are no longer alive). Although social robots are not good at understanding the needs of residents, they can be helpful in the distraction. In one study [72] we could observe that the robot was used to distract a woman who wanted to go home. In another instance, we saw that one resident did not want to do physical exercises with the care worker but was very willing to do them with the robot. So, robots have an ability to interact differently with residents and can therefore be a benefit to the daily work of care workers. Another practice of care workers is to organize group (or individual) events to activate residents. In these activities they often play games, do cognitive training, or do physical exercises. For this it is necessary to prepare and know activities. Robots can help in these activities because they can be programmed with small applications. The number of applications is almost unlimited and therefore robots can be helpful to care workers, who can simply choose an application on the robot instead of performing an activity themselves. Still, the care workers are needed to moderate the interaction, bring in the robot, start the robot or do some troubleshooting, but the task changes a somewhat and enhances the possibilities of care workers.

The third group, secondary stakeholders, is highly diverse, and it is difficult to determine general attitudes. We observed that many of those interviewed saw the technology mostly as an inevitable future. This attitude toward the robots derived from a certain degree of hopelessness. Within Germany, the status quo for care homes is very fragile in regard to the number of people working in the sector [118], and it appears unlikely to change. Indeed, it seems more likely to become worse, with ever-more people in need of care and not enough people working in this sector [19, 118, 129]. So, although the group of secondary stakeholders had ethical and data security concerns, the overall response to the social robot systems was that such systems are needed and represent a part of future care. But most saw it as something that assists the care workers in their work, not as something that takes over whole tasks. Furthermore, it was elaborated that data security must be at the core of the development because these systems could be used to manipulate residents in order to gain valuable data, but also because they could be used to observe the work of care workers; the management of a care facility could try to police the work of their employees through the system, ultimately undermining the willingness of care workers to use such systems.

Overall, we could observe that attitudes toward the robotic system were generally positive. Not all of these came from the thinking that it is a good solution; some were from the point of view that there is no other option in light of the increased pressure on the care system. Most agreed that the residents should decide whether such a system is used with them and that nobody receives diminished quality of care by electing not to use the system.

10.4 Effects of Social Robots on Daily Work Practices

The use of technology in the care sector is often connected with the viewpoint of reducing the workload of care workers or to assist in a way that enhances productivity [197]. Even though the discourse has shifted over the years toward a more practice-based approach, looking at the users and the triage between robots, residents and care workers [43, 146], prominent beliefs persist that this technology can enhance the productivity of each care worker, much like we have seen in other industries, such as car manufacturers. In modern factories, cars are produced with a high degree of involvement by robotic systems, ultimately reducing the human workforce needed in the manufacturing of the cars (but also creating new jobs, like programming robots) [78]. Technology such as robots—but others as well-are supposed to do tasks or part of tasks and in so doing alleviate the burden, or at least assist in doing the task (e.g., doing sport exercises with care home residents), thereby freeing some time for care workers. This, then, should help the care sector, which that is under constant (and growing) pressure to fulfill the work that is needed [19, 129], without adding too much budget to the sector. The goal is that fewer care workers will be needed to take care of more people at the same time. This point of view is often driven by belief in efficiency and that every sector in today's world can be optimized. But this is not the only point of view; some argue that the status-quo is already bad—that people in need of care already do not receive the care they deserve and so, after years of hoping that more people would work in this sector, that the use of technology and robots is better than nothing: that although technology is not as good as human care, at least it is something—basically acting from a position of resignation.

The long-term studies presented in this dissertation are an attempt at sensemaking of this topic; in particular, [70, 72] show that with the robot Pepper, this belief in reducing the workload cannot be sustained. The robotic system is seen as a relief by the care workers, but more in a qualitative sense. The robot can allow new ways of interacting; some residents even preferred to do physical exercises with a robotic system instead of with their human caregivers. Or, sometimes the robot is a good distraction for people with dementia in moments of crisis, helping to calm them down.

Another aspect is the preparation time. Before a social interaction with residents, social care workers must prepare what they will do with one resident or a group of residents; for example, they have to think of which physical and cognitive exercises they can do. This preparation time can be reduced with the robot, as it has such exercises installed as applications, so social care workers can choose, together with the residents on the robot, what exercises they will do. While the residents interact with the robot, such as a cognitive game application, the social care workers indeed gain some time. They are not at the center of attention because the robot is performing this task. While the application is running (run time is approximately 1–5 minutes) they can do other things. In group settings this can mean helping a resident who is struggling, filling out some documents, or (and I personally believe that this is well deserved) a short break to just watch the interaction, drink a coffee, or do nothing.

At no time did we observe in our data that the robot was left alone with the residents, something we initially thought would happen. We anticipated that care workers would be bringing the robot to the room of a resident or one of the common areas and then getting on with other things. There are several reasons why this did not happen. For one, the robot's application runs only for a few minutes, so every few minutes it is necessary to change applications, which was often done with the social care workers, who played something like a moderating role. They were bringing the robot, starting the robot, choosing the applications with the residents, and even helping the residents to interact if they did not know what they needed to do. Without this moderating role, we observed in our studies that the robot was not used. Another reason is that the residents do not completely understand how the robot work and often overtrust it, thinking that it is much "smarter" than it actually is, an effect that could also be observed in other HRI studies [148, 261, 324]. The care workers thought that this overtrust can become dangerous, as residents might think in a moment of crisis that they can rely on the robot—they cannot!) Also, care workers usually like to spend enjoyable moments with residents, and the interaction the robot creates is often an enjoyable social interaction, as residents laugh a lot and discuss in groups about what the robot is doing. Thus, another reason that the care workers did not leave the robot alone may also have been that they did not want to, that they simply enjoyed these times. Lastly, we cannot completely rule out that the care workers might have been afraid that the robot would break. We told them that the robot is insured and that they should not worry, but they still knew that it is an expensive device (€20,000) and they might have wanted to prevent it from breaking and therefore preferred to stay with it.

The robots might also trigger new job roles. The usage of technology is going to grow; as of today, technology does not play a significant role in the training of new employees. I believe it will be increasingly important to teach care workers how to use technology and to teach them its limitations. Two options can derive from it: for one, it would be possible to have some sort of technical care workers, people who have intensively studied the use of technology and that can assist other care workers with the usage of technology. The other option is to train everyone on a smaller scale. This could be done by teaching new employees within the school system and to do some on-the-job training for the existing care workers. I personally prefer the second option, as it empowers the workforce as a whole and does appear more sustainable as there is no division of skills.

But even if robotic systems could lead to a reduction of the needed workforce, if robots gain new abilitites that make them more independent, we must ask ourselves, is this something we want? How efficient do we really want to be as a society? How much do we want to go into cost-effectiveness? Do we value costs more than quality of care? Would we automate the care of our children, or do we only do that for older people in need of care? Care used to be something that is done within families and that cost families significant resources [123, 124, 254]. Nowadays we see a growing number of older people going into care homes, as family structures change and people live longer [53, 122]. To some, it might sound dystopian to bring people to a specific facility to grow old, and then to use robots to take care of them.

Societies around the world should start a debate about how much technology we want to use in care. This discussion is closely connected to the question of how we will fund the care system in the future and what an acceptable level of care might be; budgets are already tight and likely to become tighter with an increase of people in need of care due to the demographic change. There needs to be a social consensus on how this sector should operate in future. The pressure on the systems is growing and in need of sustainable solutions that have at their core the dignity of the people in need of care. We must answer the question on how efficient we want to become by using technology and what it does to the people in need of help. Although it may be the way our society works—the optimization of all aspects of life—we should also look at our way of living and the crises that arise. The way our societies are organized appear to be a poor fit for the coming future. Large crises like climate change are only beginning, so perhaps instead of further driving efficiency, we should reorganize our way of living and dedicating less energy to being productive humans and more to being caring people who give time to our loved ones.

10.5 Appropriating and Developing Social Robots in Care

Since I started writing this dissertation, around 40 applications have been installed on the robot Pepper for usage in care settings. The majority of these were developed together with the local care homes in a participatory design approach (e.g., [70, 72]). The others were open-source programs that were adapted and integrated in the system. We built a layer on the robot that allowed all applications to start directly from the robot's tablet (see 10.2).

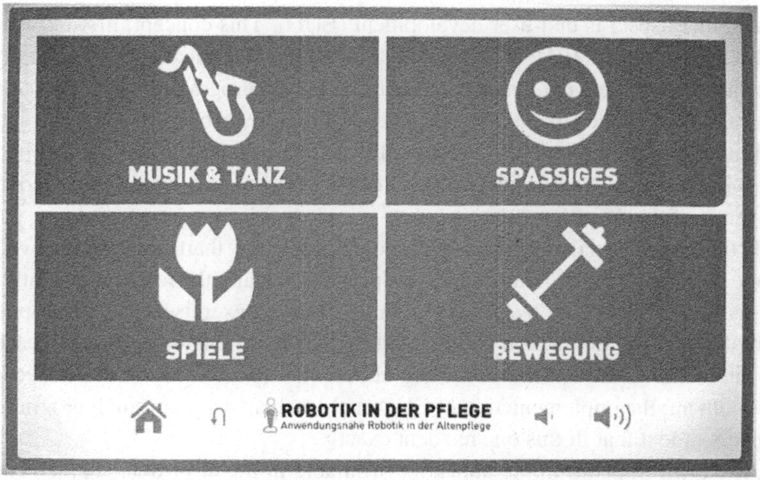

Fig. 10.2 Interface of the Robot Pepper that we developed

From 2017 to 2021, there was only one study in which we logged the usage of applications. This was because most times we, as the researchers, were present when the robot was used, so there was no knowledge to be gained by collecting such data. Also, it was important for the care homes to have as much privacy as possible. In that study [72], the robot was given to the care workers to use themselves, without the help of others, and incorporated it into their daily work. We observed that the most used applications were those that were specifically developed for the care home in a participatory design approach. For the care home, it was interesting to have software tailored to the individual needs of the participants (e.g., movement exercises with specific music or football song applications). Another aspect of local development is identification. A quiz was developed with questions about the small city where the care home was located. This made the care workers and the residents proud for several reasons. First, it was something local, which meant that the robot belonged to their community; second, the local content was developed together with the participants and, therefore, they had the feeling of having created something. We could see that this approach resulted in higher usage: despite being developed in the last half of the study, they were still among the most used applications. So, even if sometimes it is more time-consuming to tailor software to the local needs, in terms of usage it was a worthwhile investment.

Another aspect is end-user development (EUD). This concept, first introduced by Lieberman et al. [191], allows the user of the technological artifact to tailor the software themselves, not needing a technical person programming it. This allows software to be adapted to many more contexts and can even be more economical than traditional programming [340]. Often software is developed only if enough people are going to use it—that is, if it is clear that the effort of developing is exceeded by the usage of the development.

In the case of social robots in care, there are often more than three parties involved in one interaction [146]. Thus, while one person—usually the resident—is interacting with the robot (e.g., by playing a game or doing exercises), the other person is moderating the interaction and helping the resident. This other person observes the interaction and often has a feeling about what the user of the robot needs in this moment; the implementation of EUD enables them to create small programs or snippets of text that fit this one resident exactly.

As shown in detail in the appendix (available in the electronic supplementary material), the robot's applications can be separated into three categories:

- Physical Activation: Movement exercises that were designed to physically activate the residents, but also strength and relaxation training.
- Cognitive Activation: Games and exercises that help to activate memory and learning.
- Psycho-social Activation: Applications that were designed to evoke memories or to promote social interaction with the robot. These were usually made in a way to enhance users' mood and wellbeing.

We started the development of applications on a small scale with student projects or with open-source applications. Step by step, more and more applications were developed. After some time, we developed an application that had the ability to control the robot. With this application it became possible to navigate and start applications on the robot, and it was also possible to type in something that the robot would then say. In the end, more than 40 applications have been created that are accessible through an operating interface on the robot. A detailed list of all applications can be found in the appendix (available in the electronic material).

In the early studies of this work (e.g., [70, 315]), the robot was primarily used in group settings, with one robot and several residents interacting with it together. In the latest study [72], where we gave control of the robot to the care workers, we noticed that they preferred to interact with individual patients, usually in their rooms, using the robot. Normally, they would start the robot, bring it into the patient's room, and assist the patient in interacting with it. There may have been a desire for less communal activities, which could be partially explained by COVID-19.

However, we think that this was not the sole factor. The robot appears to have served as a boundary object, giving the care workers and residents something to engage together and talk about while present in close quarters. This is consistent with Kidd's findings, who noted that care workers frequently enable the interaction between resident and robot [172]. The interaction among one resident, one care worker and the robot seemed to work better for all involved in this particular care home. But at the same time, we could see that another care home prefers to use the robot with several residents at once. It appears that the robot is good for both small and bigger groups, and that it depends on the context of the situation and the individual preferences of residents and care workers.

What seems important and can also be observed in other studies (e.g., [216]) is that the appropriation was aided by talking together about the technology, whether between residents and care workers, or residents with another residents. They talk together about what works, what does not work or how they can make use of it, and sometimes they just laugh about all the limitations or unintentionally funny things the robot does. This immensely helps the process of sensemaking, to make sense of the technology and situations; it helps the users know that others also have problems and that they can discover the functionality together.

Throughout the years of our work, it became clear that there needed to be some sort of appropriation process for the care workers to use the robot in a sustainable way (specific sustainable aspects will be discussed further in the next Sect. 10.6). One study of this thesis was specifically designed for the purpose of observing the appropriation [72]. In this study, we gave the robot to one care home that had no previous experience with it. For that study we created a three-step approach: first we introduced the care workers to the technology through demonstrations and workshops and taught them how to use the robot; second, we gave them information material to look up all the necessary information themselves, like a booklet and descriptions of the applications; and third, we kept returning to the care home, mainly to observe the usage, but in the beginning we were giving a lot of support and answered all kinds of questions to the care workers, while they used the robot with the residents in their daily routines.

The study showed us that, once care workers have been properly taught to use the robot, they are more willing to implement it into their daily routines with the residents. To achieve a situation where care workers felt accustomed to the device, we taught them how to use the robot, as described above, but we also mixed in two elements of appropriation that Stevens and Pipek describe [131] as "End-User Development" (EUD) [191] and "Tailorability" [339]. EUD was achieved by giving the care worker a smartphone application that worked like a remote control for the robot and that enabled the care workers to make the robot say what they wanted or to start applications. The tailorability aspect came through the participatory design [325] of a few applications: we tailored some existing applications to the demands of the care workers and changed songs or adapted questions in the quiz game application. The tailoring of application was very successful and resulted in high usage. These results resonate with those of prior studies that used participatory design and may show that even if the process takes more time to develop, it results in usage that feels valuable to the users [186, 280].

But even if we could see that it is important to help and support in the appropriation process, it is unclear whether this can really happen in the care environment. In research projects, the participants have time and resources don't play a big role. But in the daily routine of many care homes, it is not easy to give the care workers time that is dedicated to learning, and which may be at the expense of daily routines. They are constantly needed and notoriously understaffed. What seems vital for an integration of technology such as social robots is the institutional recognition that it is important that care workers take time to learn new technologies and to give them space to try out new things, even if after the given time the result is that the technology will not be used. If the organisation is not giving enough time to the care workers, it seems unlikely that new assistive technologies will be used in a broader sense, because the potential of these will most likely not be recognized by the people who are supposed to use it.

10.6 Sustainability in Robot Usage

Sustainability[1] in research projects is a complex topic. The projects are often financed through funding agencies and are almost always limited to a few years [81, 138, 284]. That means that we, the researchers, work intensely on one topic for a short time and then run out of funding and have to do other tasks, depending on where and whether we could secure future funding. The studies in this dissertation are oriented toward action research [21, 84, 232], so we work closely with our participants and try to create a friendly atmosphere, inspecting the research artifact together. We often do our research where the participants live, which means that we go there, build bonds, accommodate participants to the technology and then leave

[1] Sustainability is often used in the context of environmental topics. In this chapter, sustainability refers to using technology long-term in an intrinsic manner by the people in the field.

afterward, once we have collected the necessary data and the founding has run out. But as others have pointed out, doing research in the field also means planning ahead of what happens once the funding runs out [229, 305]. This research practice is not sustainable; the participants and their environment get used to the robotic artifacts as well as the researchers and then, once the research is over, all activities end. Sometimes the technology can be left with the participants, but there are no updates, the developed technology (prototypes) is often prone to errors, and there is generally only limited support from the university.

In the case of the presented robotic studies in the care home, we attempted to follow another path to some degree. We worked in two care homes. In the first (care home A), we established a long-term PraxLab [8, 235]. We worked for four years with different technologies and met with the residents and the care workers in a weekly activity with different technologies (health platforms [301], exergames [317] and robots [70]). We used the same PraxLab for different research projects and were thus able to keep technology from older projects running by providing maintenance, support and fixing minor errors. But we always had to focus on the new technology, as the research projects dictated. Still, we had time to help with the older technology from previous projects.

In care home A, the robot seemed to be too complex for the care workers; after two years of using the robots together with the residents and countless contacts between care workers and the robot, we left it there to be used, but the care workers refused to use it alone. Still, there was a degree of sustainability as we, as researchers, kept on returning and using it. Later, in another study [72], we specifically taught care workers how to use the robot, did the robot become something that they used themselves without the help of the researchers. The care home is still using the robot independently, one year after we did our last study there. In the second care home (B), we took a similar approach and have different research projects. The last project is still ongoing, so it is unclear to what degree a sustainable point has been reached. The approach there was also different: we started by teaching the care workers how to use the robot.

Although some sustainability has been reached in care home A, as the care workers are using the robot without us, it is already clear that this is not a long-term solution. After some time, the robot will produce too many errors because the software is too old and not carefully enough developed, or the robot will have an accident. To really achieve sustainable usage, a company is needed that continues the R&D project and creates a product out of it [237]. A company has the possibility to create a business model out of the technology and can market it to care homes. They can then provide replacements of robots or updates to the software for a longer term than a university can, thus making it sustainable for the care home-as only then does it become possible to make a long-term commitment to the technology.

On the topic of connectedness to the existing IT infrastructure of the care home and to other healthcare providers, the research on social robots—including the presented studies of this work—is still in its infancy. The robots, more often than not, are working in silos, not connected to the main system of the care home, and are not exchanging information with other service providers. Within our studies, the robot was most often not even connected to the internet, for fear of having sensitive data exposed. But to create a meaningful usage for the robot that leads to sustainable usage, it is crucial to have such systems included in the existing system. Robot software developers must create standardized APIs that enable connection to the existing IT infrastructure. Only then can it become an integrated part. But integrating new technologies into an existing IT infrastructure goes beyond creating suitable connections, and it must also be approved by the managers of the system, as Ellen Balka points out, from experiences in another research project [29].

Apart from connectedness, it is also important to receive help from the surrounding stakeholders [206]. As Myriam Lewokowicz explains, it helps to involve politicians to get access to other big players in the research field [190], but in our experience we must also involve politicians to make them aware of the problems in the field and the regulatory changes needed to make new technologies like robots work in a sustainable way. One challenge the social robots face in the care homes in Germany is a financial one: it is difficult for a care home to spend large sums on one device—not because it would be too expensive, but because they do not have a dedicated budget for such things, even if a robot only accounts for a very small part of the whole budget. To solve this problem, political support is needed to change regulations in a way that enables technological advancements.

Lastly, it is important to reflect on the ethical side of technology developments. Often development is driven by users and their specific needs in a specific environment. But sometimes, if IT artifacts are designed by users, they may be a good fit for the local problem but in hindsight prove to be ethically problematic. To some degree, there may also be a generalizability problem: care recipients are widely diverse and live in different contexts. What is suitable for one may not be so for the other; what is useful in domestic environments may not be so in care home environments—and vice versa. An example in this is the idea to fight loneliness with social robots. Loneliness is increasingly becoming a matter of concern, and often the idea arises that lonely people in need of care should talk to robots or other IT artifacts. Although this may even work in practice and appear to create some relief, we as researchers must reflect on whether this is a good development. As Margunn Aanestad points out, we as researchers have a responsibility for the novel in this world and must assess whether an invention makes an overall good contribution in regard of dimensions like environment, society and financially [1]. But at the same time, while we try to make such assessments, we must ask ourselves whether we are competent enough to make such a decision and, when in doubt, rely on external expertise. If the general public perceives something as ethically bad, there most likely will not be sustainable integration as stakeholder advocates will withdraw their support.

10.7 Grounded Reflections and Lessons on Practice Research

During the years of creating this dissertation, many studies have been conducted and results were written down, many of them with a "lessons learned" section. But sometimes we also gained knowledge in between—knowledge that is not specific to one study and that emerged over a longer span of time, or knowledge that was unrelated to the presented study. I hereby present a collection of all the lessons I have learned. The list is long, and I try to distill only my main learning, but I believe it is incomplete—to gain a deeper understanding it is necessary to work in the field. All of the lessons are derived from working with social robots, and almost all of them in the context of older adults. I categorize the lessons I have learned in regard of methods of research, usage of the robot and the development of the robot. I believe these lessons can help future research.

10.7.1 Lessons on Development

These lessons are about the development of the social robot. They are what I have learned about how to develop the applications for the robot. I mainly draw on methods for getting information from the field to make informed design decisions.

- **Development with Locals:** Even if it is not surprising, the fact that the most frequently used applications were those that the residents or care workers of the care facility had requested indicates that we should not consider their needs and wants to be self-evident and generic. The needs in a care facility are better understood by the staff and residents than by outsiders. More crucially, insiders are much more adept at determining whether a specific application is feasible.
- **Ask the Stakeholders:** Because the structures within a care home are complex, with multiple professions working together—often in routines that have been

built up over decades—it is important to involve them in the development, to hear their opinions and constraints and to generally make them aware that something new is being developed. We must also learn from them, as their experience is firmly grounded in the daily activities of the care home.

- **Power Users:** We saw that usage of the robot is not evenly divided. Some use the robot much more than others. Those using it have a good eye for what should be further developed and are motivated to give advice. It makes sense to involve them in the development, as they are dedicating time to it as well as internally communicating the new features of the development and acting as advocates for the care home.
- **Develop with the Users in the Field:** Using participatory design and the PraxLab approach was crucial to understand the requirements, attitudes, and practices of the older adults. We learned a lot about how to develop such a robotic system and what functions it should have by working in the field and using an iterative development approach.
- **Lack of Transparency:** The robot and its algorithms are opaque to the residents and the care workers. For them, the robot is a sophisticated technological creation that they are unable to control and whose activities are occasionally difficult to account for. Future transparency may be made possible through the end-user development approach.

10.7.2 Lessons on Usage

In this section I have collected lessons about the usage or appropriation of the robot. The collection shows what I learned in the field about how usage of social robots can be enhanced. It is one thing to develop a good system, but there also needs to be good integration into the daily practices of the users.

- **Involvement of Stakeholders:** It is crucial to involve stakeholders in the process of bringing in the robot, because care is complicated and many professions deal with the same persons. It is clear that all stakeholders must be heard and informed of the process.
- **Engagement of Care Home Residents:** We observed that the older adults enjoyed interacting with the robotic system and continuously engaged with it. Routines with fixed dates were helpful. Another consideration is to make only minor incremental changes to the used applications and thus the interaction. Routines for time and behavior were smoothing factors.

- **Following Robot Instructions:** During an activity course, the residents accepted the robot as a training guide. They were willing to follow the robot's instructions in physical training and gaming situations, although human moderation was occasionally required.
- **Integration with Local Activities:** Recognizing existing structures, roles and duties, time frames and routines is crucial to determining feasibility. In one of our studies, the robot was incorporated into the weekly plan so that the residents and care workers were informed of when and how the robot would be used. We contend that the integration process should be created by the facility itself rather than being imposed hierarchically.
- **Voluntariness:** The voluntariness principle appears to be significant in terms of robot use. The worth of the robot in actual use depended greatly on how much the individual workers or residents found it valuable and relevant. Additionally, the individual's excitement, willingness, and dedication to using the technology were crucial to its adoption.
- **Technical Support:** Although technical help will always be required, how much of it is required will depend on how willing care workers are to handle issues and acknowledge that technical failures can occur, and that they are typically not their fault and shouldn't be attributed to them. Such issues affect both individuals and organizations. They rely on a healthy understanding among all parties that problems can emerge. Technical support helps to overcome problems that seem to difficult and should be easily available; in the best case, it can be provided over the phone.
- **Power Users:** We saw that certain staff members acted as "power users," advancing the project, supporting one another, and taking charge if issues arose. A certain level of "enskilling" would improve the social caregiver's job and also improve the robot's potential role.
- **Quality of Care Instead of Workload Reduction:** The robot was viewed as a helpful addition to the regular tasks of care workers. However, there were not significant savings in terms of labor or time. Benefits are more closely related to fresh opportunities for providing an enhanced care.
- **Trust Agents:** The participants' trust of the robot has increased as a result of including care workers whom they knew and frequently interacted with. That the care workers trusted the researchers helped immensely, as the residents used the care workers' opinion of the researchers to make sense of the situation.
- **Human-Robot Support:** It takes a moderator to explicitly facilitate engagement with the robot; occasionally, the apps' functioning has to be clarified, the robot needs to be relocated from one location to another, or someone needs to inspire the participants to engage with the system.

10.7.3 Lessons on Methods

The lessons on methods are about how I approached the field and how I made sense of it. I have learned a few things that helped me to better understand the field, and I hope that others can use some of the knowledge I gained and appropriate it. Because my dissertation is based mainly on qualitative research results, the lessons I present here are made for this. Other research paradigms might draw other lessons.

- **Qualitative User Research:** The appropriate amount of information to develop social robots in care settings is provided by qualitative methods with participatory design. They are crucial to comprehend the complicated interactions between humans and robots. Robot development can be carried out in a delicate and nuanced manner by expanding on these.
- **Wizard-of-Oz to Understand the Interaction:** Researchers can better understand a robot's perspective and potential roles by using Wizard-of-Oz techniques, wherein a human controls the robot and interacts with people to gain a unique understanding of how people interact with robots-and, to some degree, incorporate it. Expanding on the Wizard-of-Oz methods helps to understand the role of the robot and how it should behave in certain situations.
- **Develop Trustful Relationships with Participants:** With the PraxLab approach, researcher and participants look together at the robot and develop it together. They work closely together—sometimes for years—and must be open with each other. Participants tend to want to give something back to the researchers and thus taint the feedback on the robot by skewing it toward the positive. To avoid this, it is important to build a trustful relationship and to make clear that all results are good ones. This yields results that are more robust.
- **Take the Blame:** If something does not work, it is crucial to accept the blame or to blame the technology, instead of blaming the participant-even if it was their fault. The first priority should be that participants are open toward the technology; blaming the participant only results in insecurities about usage, further complicating the creation of future results.
- **Replacement of Participants:** Participants do not always stay until the end of a study; the reasons for this are manifold. The drop-out rate due to health issues is elevated in care home residents, compared to other participant groups, as the age and health risks are high. It is helpful to create a study design that from the beginning is aware of these risks and has strategies to onboard new participants.
- **Sustainable Integration:** Consider sustainability from the beginning. Universities may find it difficult to achieve long-term integration of a developed product because research projects end after a few years. Participants often no longer

have access to developed products once a project has run its course because the products are not offered on the market. Before a project starts, it is important to establish who has the rights to future development and how its results will be maintained in the field.

- **Appropriation in Groups:** Involvement with the robot has been significantly impacted by the group environment. Residents learned from one another and relied on one another to interpret the actions of the robots.

11 Limitations

The limitations here take a broader approach, as they are already specified within the publication of this dissertation. In the following I will describe limitations that were unspecific to a single publication but were instead reoccurring.

One reoccurring limitation is that because all participants were asked whether they wanted to work with us, only people who were interested in our work participated. People who were not interested normally did not want to work with us. (There was no compensation for participating.) Still, we encouraged all participants to work with us, and some who were not interested did, as we were telling them that all opinions are important. Some worked with us because it was important for them to help the research, but they were outnumbered by people who were genuinely interested in the technology.

Another limitation is that this work was done mainly with qualitative research methods; some parts were quantitative, but at no time were representative study designs used. Accordingly, this work has an explanatory focus and goes in depth into interaction and usage in real-world scenarios. Further studies are needed to control for whether the presented results have a representative character.

Due to our approach to work in the field with a long-term perspective, we built relationships with the participants and other stakeholders in the field. This led to a mixed role: we were not only researchers but also acted on intentions that were not research-oriented but had a friendly or helpful motivation. This may have tainted the relationships we had with the participants and probably had an effect on the data that was created. This behavior certainly created more trust among the participants toward us, but it also may have resulted in positively tainted data, as the participants might also have tried to do us a favor-even though we always said that negative results are good results.

F. Carros, *Design, Development and Sensemaking of Human-Robot Interaction in Care Settings*, https://doi.org/10.1007/978-3-658-45233-9_11

Another limitation is that not everything that happened was recorded. Due to our long-term work, we were simply not able to write everything down or record everything. Sometimes there were also situations that were highly private or sensitive; such data was not recorded and not used. It nevertheless helped me to understand the field.

Secondary knowledge was sometimes used to make sense of the observations. Although residents were always the biggest group of interviewees, we learned a lot from social care workers. They knew how to explain the behavior and attitudes of the residents, almost as if they were advocates for them. This, of course, taints results in a direction that largely represents their views or wants. But when working with residents, we rarely received a lot of information; residents were often exhausted by the interviews and tended to give short answers, rarely reflecting on the past interactions.

Although we tried to be as open as possible to the needs of the field, we were also bound to the existing technology. We did not develop new robots; rather, we developed software for robots that we bought on the open market. This narrowed the possibilities of tackling the needs of the field. We could only develop what was technically feasible with the robot. The scenario of a robot bringing drinks to residents is one such limitation: the social robots we worked with lacked the capability to transport things with their "hands," so we could not fulfill this need with the given system. Although it was frequently requested in the participatory design sessions.

Lastly, major parts of this dissertation were created during the COVID-19 pandemic; the results might have been different otherwise. I'm personally happy that it was possible to continue the work, but I must admit that it was different than it would have been under normal conditions. Consequently, parts of the results are probably unique to the given situation. Some interviews were done in video calls. Often the residents needed help with the video call devices, which resulted in care workers helping them and staying with them during the interviews. The resident may have not said certain things because a third person was listening.

12 Conclusion and Outlook

In the final chapter of this dissertation, I will outline what main findings have resulted from my work, what I have learned in this intense time in regard to methods, technology, and societal trends. I will further elaborate on what may be important for future work.

12.1 Conclusion on our Findings

The results I and my colleagues have achieved demonstrate (given the limitations) that social robots have a place in a care homes, and residents have shown that they enjoy interacting with them. The robots engage in a range of activities, including games as well as physical and cognitive training. Different robot models have their own logic and advantages, but the robots are all alike in that they activate and stimulate the residents and are, therefore, positive influences on their wellbeing.

Perhaps the most important lesson we learned was that the robots do not work alone. A person is needed to take care of the robot and manage its activities, whether to start it, bring it from one room to another, or act as a moderator. The use of robots, we discovered, allows workers to take small breaks—but otherwise they had little effect on workload. The wholesale substitution of robots for care workers is, for the moment, a fairy tale. Social robots, however, can be of assistance in some practices of care workers, especially social care workers. We have seen in later studies that care workers are willing to learn how to use the robot and that they do use it in their daily practices. There is no need for a technician at all times; it is enough to offer support, build robust software, and accommodate the process of appropriation.

At the core of our development and integration of social robots in care homes was the element of participatory design [325]. Not only is it important to make developments that are oriented to the needs of the people in the field, but I also

F. Carros, *Design, Development and Sensemaking of Human-Robot Interaction in Care Settings*, https://doi.org/10.1007/978-3-658-45233-9_12

believe it is important that this vulnerable user group is heard to build common trust and understanding. I believe this participatory approach increased the quality of our work over time. Residents understood that our interest is in their wellbeing, and care workers understood that we did not intend to replace them, but rather aimed to provide what was, above all, an assistive technology. Trust also helped to establish a fruitful relationship. The participants in the field sometimes wanted to be kind to the researchers and, at least in the short term, their answers had to be treated with caution. With the development of a trustful relationship, it became easier to make the participants understand that negative results are important results as well.

We learned that care homes, and care in general, is a multi-stakeholder environment with many different interest groups: residents, families, care workers, health insurance companies, society, care home providers, and more. It was important, therefore, to understand the different roles that stakeholders played and to involve them in developments. It is often not possible to involve all stakeholders to the same degree—for reasons such as time, willingness, and cognitive ability—but it was important, nonetheless, to have an open view of possibilities for communication and wider participation. The development and appropriation of social robots in care settings over the span of four years has shown that a long-term approach enables a shared vision and trustful work relationship. It was important that all stakeholders came to know us and our intentions. The residents often needed time to adapt their routines to our experiments. Tests with the robot became better with time, as the residents and the stakeholders surrounding them got used to us and the activities associated with the social robot.

Sustainability, in the sense of making technology that lasts, emerged as a topic. We engaged the residents and the care home and got them used to our robot developments. We wanted them to immerse themselves in this development to see how they use it long-term. It is only fair to find solutions for how these robot developments can remain with them after the research is done. And lastly, we must adapt to the dynamics of the field. We influenced the care home, and they influenced us. Every invention brings a new dynamic. No situation is totally static, and they must always be contextualized.

12.2 Conclusion on the Field

My feeling, admittedly personal, is that conditions in the care sector are becoming worse (which is also indicated by statistics [56]). In 2018, when we started the research, care workers were sometimes reluctant to use social robots. Some were critical about data security, others about the ethics, and some just preferred to not use technology in their daily work. Another topic that they raised was their fear of being replaced by a robot, as other sectors in industry have seen in the last years (even though it was more a shift to different occupations than a loss of jobs [4, 46, 189]). Today, a few years and one pandemic later, these attitudes have changed. Nowadays, if we come with a new idea or a new robot to the care home, they are usually very open. I found that frequently they need everything that could possibly help. I don't have the feeling that people have hope that the system will change for the better (e.g., better pay or more people managing and caring for the same number of residents). Many care workers are burned out, and the government has shown again and again that they are not willing (or able) to create a better working environment. Some gradual changes (like small increases in pay) have occurred, but the sector is in need of more drastic measures. Therefore, we observe an increase in openness toward technology to solve the problem. We have seen that technology, like robots, can be helpful tools, but its use fundamentally depends on the involvement of care workers. Robots can help, but as of today, they will not solve the problem. As for a societal evaluation of social robots in care settings, we have involved citizens in our research (see [68]), and this experience has taught us that the majority of the participants were not strongly opposed to the use of robots in such environments, if carefully developed and as an assistance to the actual workers. Many see robotic assistance as an added benefit for the sector and are open to it.

There are a number of problems that seem to be worsening and will likely have effects on the stability of the care system. The demographic change in the Western Hemisphere is ongoing and we will see a growing number of people in need of care [42]. However, the number of people willing to work in this sector is not growing at the same pace. Furthermore, it seems that family structures are changing in comparison to the standard pattern of the Baby Boomer generation. The model of one person staying at home and caring for the parents is becoming more difficult to sustain as these people are needed in the workforce and households often cannot afford to live on one income, as it used to be.

For now, richer countries like Germany are attracting care workers from other, less wealthy countries to reduce the pressure on the care system. This is possible because they can pay them better or can provide a secure social environment. This may change in future, with an aging population and the increase of crises (e.g., climate crisis, wealth inequality). It is unclear whether countries with aging societies can remain attractive to foreign workforces. The demographic change means that fewer people will be working, while at the same time, more people will be in need of help. This may diminish the wealth of a country and its reactivity to crisis, ultimately resulting in a reduced pull effect for foreign workers. Countries with an aging population should not be too certain that they will be able to replace their shortfalls with foreign workers and must look into alternative solutions that are sustainable and enduring.

12.3 Outlook on Future Research

The experience of recent years and familiarity with the work of other researchers (including in other disciplines) inclines me to believe that the sector of "social robots in care settings" will remain an important topic.

As described, the challenges for the sector are manifold, and my personal outlook is rather pessimistic. This macro trend means either that the system needs greater assistance or that the system needs reforming to something that requires fewer people working in this sector. I think that in both scenarios, the possibilities of technology are growing in significance.

A number of developments seem promising. One is the collaboration of different robots and digital systems. We often focus on individual digital devices, but I think it makes sense to look more at how the digital systems can collaborate. One system is often insufficient in some aspects while other systems are insufficient in other dimensions. I believe it makes sense to investigate how technologies can complement one another to compensate for the individual deficits each might have. We should not try to reinvent the human care worker, but we should try to develop systems that work together and that also factor in the human. Machines will help the residents, the care workers, and the other machines.

Another element is artificial intelligence (AI), which has already been used in the developments discussed in this dissertation. The topic of AI has been highly discussed in the last years and it seems to be ongoing. I expect a continuously high amount of investment in this sector and commensurate advances to follow. With AI, the machine vision functions and the "natural language" dialog systems become interesting for social robots in care. I believe we will see some developments that will have a value for social robots in care homes and the potential to advance the field.

In my experience, future developments should be achieved through participatory design-designing with the people as well as for the people. We have already seen too many developments that were fancy but in the end not used. Sensitive development should become a standard procedure: to first understand the field and the practices of the field, and then to develop.

Although I see many opportunities in the use of robotic technology, there are also challenges on the horizon. With the increasing delegation of tasks to technology, we need to involve stakeholders and ethical awareness more in the development.

Although humans usually have values, robotic machines never do. Robots that may be used to maximize efficiency will not question the tasks they are instructed to do. As humans, however, we are entitled to ask whether efficiency is the only measure of success. Care is not about efficiency—it entails, instead, judgements about "quality": what is good for those vulnerable people in our society who are in need of care (and for those who give care). Although some technological assistance is certainly needed, given the circumstances, we also need to create a common understanding of what we want as a society. How such an understanding can be reached remains an open question. The calculation of cost and benefit will be, as it already is, a highly complex matter that inevitably involves judgements about wellbeing. Perhaps it is time to evaluate whether we are on the right path. Robot technology will be increasingly used in the care sector because of demographic change and the change in family structures. How it will be used, however, is a different—and more difficult—matter. Because we so rarely have the time to care for our loved ones, we increasingly delegate this task to care homes. Although in some sense this has a liberating aspect—as we are no longer obligated to provide care—we should think about what we do with this time gained. Perhaps we should be dedicating more time to our community and the environment instead of trying to keep alive an already unsustainable economy.

Bibliography

1. Margunn Aanestad. „Sustainability and user-driven IT projects“. In: *Future-Proofing: Making Practice-based IT Design Sustainable*. Oxford University Press, 2022.
2. Jordan Abdi, Ahmed Al-Hindawi, Tiffany Ng, and Marcela P Vizcaychipi. „Scoping review on the use of socially assistive robot technology in elderly care“. In: *BMJ open* 8.2 (2018), e018815.
3. Gregory D Abowd, Aaron F Bobick, Irfan A Essa, Elizabeth D Mynatt, and Wendy A Rogers. „The aware home: A living laboratory for technologies for successful aging“. In: *Proceedings of the AAAI-02 Workshop "Automation as Caregiver*. 2002, pp. 1–7.
4. Daron Acemoglu and Pascual Restrepo. „The race between man and machine: Implications of technology for growth, factor shares, and employment“. In: *American Economic Review* 108.6 (2018), pp. 1488–1542.
5. Evan Ackerman. *The Man Who Coined the Word "Robot" Defends Himself.* https://spectrum.ieee.org/karel-capek-robots. Accessed: 08.02.2024. 2024.
6. Michael Agar and James MacDonald. „Focus groups and ethnography“. In: *Human organization* 54.1 (1995), pp. 78–86.
7. European Environment Agency. *European Environment Agency Marine LitterWatch*. n. d.
8. Michael Ahmadi, Anne Weibert, Corinna Ogonowski, et al. „Challenges and lessons learned by applying living labs in gender and IT contexts“. In: *Proceedings of the 4th Conference on Gender & IT*. 2018, pp. 239–249.
9. Paul Aktinson and Martyn Hammersley. „Ethnography and participant observation“. In: *Strategies of Qualitative Inquiry. Thousand Oaks: Sage* (1998), pp. 248–261.
10. Andy Alaszewski. *Using diaries for social research*. Sage, 2006.
11. Laura Albornos-Muñoz, Marìa Teresa Moreno-Casbas, Clara Sánchez-Pablo, et al. „Efficacy of the Otago Exercise Programme to reduce falls in community-dwelling adults aged 65–80 years old when delivered as group or individual training“. In: *Journal of advanced nursing* 74.7 (2018), pp. 1700–1711.
12. Bianca Ambrose-Oji, Alexander PN van der Jagt, and Sue O'Neil. „Citizen Science: Social Media as a supporting tool“. In: *Farnham: Forest Reseach* (2014).
13. Robotics Institute of America. „News“. In: *Robotics Today* (1980).
14. Elizabeth L. Anderson, Paul Turnham, John R. Griffin, and Chester C. Clarke. „Consideration of the aerosol transmission for COVID-19 and public health“. In: *Risk Analysis* 40.5 (2020), pp. 902–907.

F. Carros, *Design, Development and Sensemaking of Human-Robot Interaction in Care Settings*, https://doi.org/10.1007/978-3-658-45233-9

15. Michael Anderson and Susan Leigh Anderson. „Robot be good“. In: *Scientific American* 303.4 (2010), pp. 72–77.
16. Rebecca Andreasson, Beatrice Alenljung, Erik Billing, and Robert Lowe. „Affective Touch in Human–Robot Interaction: Conveying Emotion to the Nao Robot“. In: *Int. J. Social Rob.* 10.4 (Sept. 2018), pp. 473–491.
17. Antonio Carnevale. „Robots, Disability, and Good Human Life“. In: *Disability Studies Quarterly* 35.1 (2015).
18. Statistik der Bundesagentur für Arbeit. „Blickpunkt Arbeitsmarkt—Akademikerinnen und Akademiker“. In: *Bundesagentur für Arbeit: Nürnberg, Germany* (2019).
19. Bundesagentur für Arbeit Statistik/Arbeitsmarktberichterstattung. *Arbeitsmarktsituation im Pflegebereich*. 2022.
20. European Citizen Science Association. *European Citizen Science Association*. n. d.
21. David E Avison, Francis Lau, Michael D Myers, and Peter Axel Nielsen. „Action research“. In: *Communications of the ACM* 42.1 (1999), pp. 94–97.
22. Shiri Azenkot, Catherine Feng, and Maya Cakmak. „Enabling building service robots to guide blind people a participatory design approach“. In: *2016 11th ACM/IEEE International Conference on Human-Robot Interaction (HRI)*. IEEE. 2016, pp. 3–10.
23. Nicholas Babchuk. „The role of the researcher as participant observer and participant-as-observer in the field situation“. In: *Human organization* 21.3 (1962), pp. 225–228.
24. Matthias Bäcker. „In J. Kühling & B. Buchner“. In: *Datenschutz-Grundverordnung BDSG Kommentar* (2018).
25. Stefanie Baisch, Thorsten Kolling, Saskia Rühl, et al. „Emotional robots in a nursing context : Empirical analysis of the present use and the effects of Paro and Pleo“. In: *Z. Gerontol. Geriatr.* 51.1 (2018), pp. 16–24.
26. Stefanie Baisch, Thorsten Kolling, Arthur Schall, et al. „Acceptance of social robots by elder people: does psychosocial functioning matter?“ In: *International Journal of Social Robotics* 9.2 (2017), pp. 293–307.
27. Markus Bajones, David Fischinger, Astrid Weiss, Daniel Wolf, and Susanne Frennert. „Hobbit: Providing Fall Detection and Prevention for the Elderly in the Real World“. In: *J. Rob.* 2018 (2018), pp. 1–20.
28. Lynda Baker. „Observation: A complex research method“. In: *Library trends* 55.1 (2006), pp. 171–189.
29. Ellen Balka. „Persistence and patience and relations—ActionADE at eight“. In: *Future-Proofing: Making Practice-based IT Design Sustainable*. Oxford University Press, 2022.
30. Ellen Balka and Ina Wagner. „Making things work: dimensions of configurability as appropriation work“. In: *CSCW '06: Proceedings of the 2006 20th anniversary conference on Computer supported cooperative work*. New York, NY, USA: Association for Computing Machinery, Nov. 2006, pp. 229–238.
31. Kimberly A. Barchard, Leiszle Lapping-Carr, R. Shane Westfall, et al. „Measuring the Perceived Social Intelligence of Robots“. In: *J. Hum.-Robot Interact.* 9.4 (2020).
32. Christoph Bartneck, Tony Belpaeme, Friederike Eyssel, et al. „What is Human-Robot Interaction?“ In: *Human-robot interaction: An introduction*. Cambridge University Press, 2020.
33. Heidrun Becker. „Robotik in der Gesundheitsversorgung: Hoffnungen, Befürchtungen und Akzeptanz aus Sicht der Nutzerinnen und Nutzer“. In: *Pflegeroboter*. Springer Gabler, Wiesbaden, 2018, pp. 229–248.

34. Roger Bemelmans, Gert Jan Gelderblom, Pieter Jonker, and Luc de Witte. „Socially assistive robots in elderly care: a systematic review into effects and effectiveness". In: *J. Am. Med. Dir. Assoc.* 13.2 (2012), pp. 114–1201.
35. Till Bieg, Cornelia Gerdenitsch, Isabel Schwaninger, Bettina Manuela Johanna Kern, and Christopher Frauenberger. „Evaluating Active and Assisted Living technologies: Critical methodological reflections based on a longitudinal randomized controlled trial". In: *Computers in Human Behavior* 133 (Aug. 2022), p. 107249.
36. Deborah R. Billings, Kristin E. Schaefer, Jessie Y.C. Chen, and Peter A. Hancock. „Human-Robot Interaction: Developing Trust in Robots". In: *Proceedings of the Seventh Annual ACM/IEEE International Conference on Human-Robot Interaction.* HRI '12. Boston, Massachusetts, USA: Association for Computing Machinery, 2012, pp. 109–110.
37. Birgit Kramer. „Dementia Caregivers in Germany and Their Acceptance of New Technologies for Care: The Information Gap: Figure 1". In: *Public Policy & Aging Report* 24.1 (2013), pp. 32–34.
38. Gro Bjerknes, Pelle Ehn, and Morten Kyng. „Computers and Democracy: A Scandinavian Challenge. Aldershot". In: *UK: Gower* (1987).
39. Alexis E. Block and Katherine J. Kuchenbecker. „Softness, warmth, and responsiveness improve robot hugs". In: *International Journal of Social Robotics* 11.1 (2019), pp. 49–64.
40. Jeanette Blomberg and Helena Karasti. „Ethnography: Positioning ethnography within participatory design". In: *Routledge international handbook of participatory design.* Routledge, 2012, pp. 86–116.
41. Jeanette L. Blomberg and Austin Henderson. „Reflections on Participatory Design: Lessons from the Trillium Experience". In: *Proceedings of the SIGCHI Conference on Human Factors in Computing Systems.* CHI '90. New York, NY, USA: Association for Computing Machinery, 1990, pp. 353–360.
42. Karl Blum, Matthias Offermanns, and Petra Steffen. „Situation und Entwicklung der Pflege bis 2030". In: *Düsseldorf: Deutsches Krankenhausinstitut e* 2019 (2019).
43. BMFSFJ. *Achter Altersbericht. Ä ltere Menschen und Digitalisierung.* Ed. by Bundesministerium für Familie, Senioren, Frauen und Jugend. Berlin, 2020.
44. Patrick Boissy, Hélène Corriveau, François Michaud, Daniel Labonté, and Marie-Pier Royer. „A qualitative study of in-home robotic telepresence for home care of community-living elderly subjects". In: *Journal of telemedicine and telecare* 13.2 (2007), pp. 79–84.
45. Aletta Bonn and Katrin Vohland. „Citizen science strategie 2020 für Deutschland". In: *Online verfügbar unter* https://www.buergerschaffenwissen.de/sites/default/files/assets/dokumente/gewiss-gruenbuchcitizensciencestrategie.pd 25.11.2022 (2016).
46. Jeff Borland and Michael Coelli. „Are robots taking our jobs?" In: *Australian Economic Review* 50.4 (2017), pp. 377–397.
47. Jeffrey Bradshaw, Paul J. Feltovich, and Matthew Johnson. „Human-Agent Interaction". In: *Handbook of Human-Machine Interaction* (Jan. 2011), p. 20.
48. Virginia Braun and Victoria Clarke. „Using thematic analysis in psychology". In: *Qualitative research in psychology* 3.2 (2006), pp. 77–101.

49. Virginia Braun, Victoria Clarke, Nikki Hayfield, and Gareth Terry. „Thematic Analysis". In: *Handbook of Research Methods in Health Social Sciences*. Ed. by Pranee Liamputtong. Singapore: Springer Singapore, 2018, pp. 1–18.
50. Cynthia Breazeal. *Designing sociable robots*. MIT press, 2004.
51. Elizabeth Broadbent, Rebecca Stafford, and Bruce MacDonald. „Acceptance of healthcare robots for the older population: Review and future directions". In: *International journal of social robotics* 1.4 (2009), pp. 319–330.
52. Joost Broekens, Marcel Heerink, Henk Rosendal, et al. „Assistive social robots in elderly care: a review". In: *Gerontechnology* 8.2 (2009), pp. 94–103.
53. Marjolein I Broese van Groenou and Alice De Boer. „Providing informal care in a changing society". In: *European Journal of Ageing* 13.3 (2016), pp. 271–279.
54. Statistisches Bundesamt. *Bevölkerung im Wandel. Annahmen und Ergebnisse der 14. Koordinierten Bevölkerungsvorausberechnung*. 2019.
55. Statistisches Bundesamt. *Mikrozensus 2013—Sonderauswertungen för das Deutsche Zentrum für Altersfragen*. 2015.
56. Bundesgesundheitsministerium. *Pflegeversicherung - Leistungsempfänger der sozialen Pflegeversicherung am Jahresende nach Pflegestufen 1995–2021*.
57. Bundesministerium. *Bundesministerium für Bildung und Forschung. Roboter für Assistenzfunktionen Interaktionsstrategien*. n. d.
58. A. A. Burstein, Olivia DaDalt, Birgit Kramer, Lisa A. D'Ambrosio, and Joseph F. Coughlin. „Dementia caregivers and technology acceptance: interest outstrips awareness". In: *Gerontechnology* 14.1 (2015), pp. 45–56.
59. Maurizio Bussolo, Johannes Koettl, and Emily Sinnott. *Golden aging: Prospects for healthy, active, and prosperous aging in Europe and Central Asia*. World Bank Publications, 2015.
60. John-John Cabibihan, Hifza Javed, Marcelo Ang, and Sharifah Mariam Aljunied. „Why robots? A survey on the roles and benefits of social robots in the therapy of children with autism". In: *International journal of social robotics* 5.4 (2013), pp. 593–618.
61. Rafael A. Calvo, Dorian Peters, Daniel Johnson, and Yvonne Rogers. „Autonomy in technology design". In: *CHI EA '14: CHI '14 Extended Abstracts on Human Factors in Computing Systems*. New York, NY, USA: Association for Computing Machinery, Apr. 2014, pp. 37–40.
62. Cambridge Business English Dictionary. *"robot"*. Accessed: 2022-05-27.
63. J Capek. „Opilec". In: *Lelio A Pro Delfina. Aventinum, Prague* (1925).
64. Karel Capek. *RUR and the Vision of Artificial Life*. MIT Press, 2024.
65. Karel Capek. „The meaning of RUR". In: *Saturday Rev* 136 (1923), p. 79.
66. Felix Carros. „Roboter in der Pflege, ein Schreckgespenst?" In: *Mensch und Computer 2019—Workshopband*. Bonn: Gesellschaft für Informatik e.V, 2019.
67. Felix Carros, Hannes Eilers, Johanna Langendorf, et al. „Roboter als intelligente Assistenten in Betreuung und Pflege—Grenzen und Perspektiven im Praxiseinsatz". In: *Künstliche Intelligenz im Gesundheitswesen*. Springer, 2022, pp. 793–819.
68. Felix Carros, Johanna Langendorf, Dave Randall, Rainer Wieching, and Volker Wulf. „Citizen Participation in Social Robotics Research". In: *Meaningful Futures with Robots—Designing a New Coexistence*. Chapman and Hall/CRC, 2022, pp. 205–216.
69. Felix Carros, Johanna Meurer, Diana Löffler, et al. „Exploring Human-Robot Interaction with the Elderly: Results from a Ten-Week Case Study in a Care Home".

In: *Proceedings of the 2020 CHI Conference on Human Factors in Computing Systems*. CHI '20. Honolulu, HI, USA: Association for Computing Machinery, 2020, pp. 1–12.

70. Felix Carros, Johanna Meurer, Diana Löffler, et al. „Exploring human-robot interaction with the elderly: results from a ten-week case study in a care home". In: *Proceedings of the 2020 CHI Conference on Human Factors in Computing Systems*. 2020, pp. 1–12.
71. Felix Carros, Isabel Schwaninger, Adrian Preussner, et al. „Care Workers making Use of Robots: Results of a 3 Month Study on Human-Robot-Interaction within a Care Home". In: *Proceedings of the 2022 CHI Conference on Human Factors in Computing Systems*. New Orleans, LA, USA: ACM, 2022.
72. Felix Carros, Isabel Schwaninger, Adrian Preussner, et al. „Care Workers Making Use of Robots: Results of a Three-Month Study on Human-Robot Interaction within a Care Home". In: *Proceedings of the 2022 CHI Conference on Human Factors in Computing Systems*. CHI '22. New Orleans, LA, USA: Association for Computing Machinery, 2022.
73. Felix Carros, Tobias Störzinger, Anne Wierling, Adrian Preussner, and Peter Tolmie. „Ethical, Legal Participatory Concerns in the Development of Human-Robot Interaction: Lessons from Eight Research Projects with Social Robots in Real-World Scenarios". In: *i-com* 21.2 (2022), pp. 299–309.
74. Nico Castelli, Corinna Ogonowski, Timo Jakobi, et al. „What Happened in My Home? An End-User Development Approach for Smart Home Data Visualization". In: *Proceedings of the 2017 CHI Conference on Human Factors in Computing Systems*. CHI '17. Denver, Colorado, USA: Association for Computing Machinery, 2017, pp. 853–866.
75. Débora de Castro Leal, Max Krüger, Kaoru Misaki, David Randall, and Volker Wulf. „Guerilla Warfare and the Use of New (and some old) Technology: Lessons from FARC's Armed Struggle in Colombia". In: *Proceedings of the 2019 CHI conference on human factors in computing systems*. 2019, pp. 1–12.
76. Jim X. Chen. „The Evolution of Computing: AlphaGo". In: *Computing in Science & Engineering* 18.4 (2016), pp. 4–7.
77. Ed H Chi. „A position paper on'living laboratories': Rethinking ecological designs and experimentation in human-computer interaction". In: *International Conference on Human-Computer Interaction*. Springer. 2009, pp. 597–605.
78. VC Chijindu and HC Inyiama. „Social implications of robots–An overview". In: *International Journal of Physical Sciences* 7.8 (2012), pp. 1270–1275.
79. Looi Theam Choy. „The strengths and weaknesses of research methodology: Comparison and complimentary between qualitative and quantitative approaches". In: *IOSR journal of humanities and social science* 19.4 (2014), pp. 99–104.
80. Anders BH Christensen, Christian R Dam, Corentin Rasle, et al. „Reducing overtrust in failing robotic systems". In: *2019 14th ACM/IEEE International Conference on Human-Robot Interaction (HRI)*. IEEE. 2019, pp. 542–543.
81. Lutz Claes, Edmund A.M. Neugebauer, Peter Biberthaler, and Ute Schäfer. „Projektplanung und Projektantrag". In: *Von der Idee zur Publikation: Erfolgreiches wissenschaftliches Arbeiten in der medizinischen Forschung*. Ed. by Edmund A.M Neugebauer, Wolf Mutschler, and Lutz Claes. Berlin, Heidelberg: Springer Berlin Heidelberg, 2011, pp. 47–77.
82. Max E Clarkson. „A stakeholder framework for analyzing and evaluating corporate social performance". In: *Academy of management review* 20.1 (1995), pp. 92–117.

83. Sebastian Cmentowski and Jens Krüger. „Playing With Friends—The Importance of Social Play During the COVID-19 Pandemic". In: *Extended Abstracts of the 2020 Annual Symposium on Computer-Human Interaction in Play*. CHI PLAY '20. Virtual Event, Canada: Association for Computing Machinery, 2020, pp. 209–212.
84. Louis Cohen, Lawrence Manion, and Keith Morrison. „Action research". In: *Research methods in education*. Routledge, 2017, pp. 440–456.
85. European Commission. *European Commission Citizen Science*. n. d.
86. Torbjørn S Dahl and Maged N Kamel Boulos. „Robots in health and social care: A complementary technology to home care and telehealthcare?" In: *Robotics* 3.1 (2013), pp. 1–21.
87. Nils Dahlbäck, Arne Jönsson, and Lars Ahrenberg. „Wizard of Oz studies—why and how". In: *Knowledge-based systems* 6.4 (1993), pp. 258–266.
88. Karin Dahlberg. „The enigmatic phenomenon of loneliness". In: *International Journal of Qualitative Studies on Health and Well-being* 2.4 (Jan. 2007), pp. 195–207.
89. Christine Dearnley. „A reflection on the use of semi-structured interviews". In: *Nurse researcher* 13.1 (2005).
90. Michael Decker, Mathias Gutmann, and Julia Knifka. *Evolutionary Robotics, Organic Computing and Adaptive Ambience: Epistemological and Ethical Implications of Technomorphic Descriptions of Technologies*. Lit, 2015.
91. Jan Derboven, Roos Voorend, and Karin Slegers. „Design trade-offs in self-management technology: the HeartMan case". In: *Behaviour & Information Technology* (July 2019).
92. Dmitry Dereshev, David Kirk, Kohei Matsumura, and Toshiyuki Maeda. „Long-Term Value of Social Robots through the Eyes of Expert Users". In: *CHI '19: Proceedings of the 2019 CHI Conference on Human Factors in Computing Systems*. New York, NY, USA: Association for Computing Machinery, 2019, pp. 1–12.
93. Ann DeSmet, Debbe Thompson, Tom Baranowski, et al. „Is participatory design associated with the effectiveness of serious digital games for healthy lifestyle promotion? A meta-analysis". In: *Journal of medical Internet research* 18.4 (2016), e4444.
94. Carsten Dochow. „Telemedizin und Datenschutz". In: *Medizinrecht* 37.8 (2019), pp. 636–648.
95. Steven Dow, Blair MacIntyre, Jaemin Lee, et al. „Wizard of Oz support throughout an iterative design process". In: *IEEE Pervasive Computing* 4.4 (2005), pp. 18–26.
96. Janice J Eng. „Fitness and mobility exercise (FAME) program for stroke". In: *Topics in geriatric rehabilitation* 26.4 (2010), p. 310.
97. Mats Eriksson, Veli-Pekka Niitamo, Seija Kulkki, et al. „State-of-the-art in utilizing Living Labs approach to user-centric ICT innovation-a European approach". In: *Lulea: Center for Distance-spanning Technology. Lulea University of Technology Sweden: Lulea* (2005).
98. Connor Esterwood, Kyle Essenmacher, Han Yang, Fanpan Zeng, and Lionel Peter Robert. „A Meta-Analysis of Human Personality and Robot Acceptance in Human-Robot Interaction". In: *Proceedings of the 2021 CHI Conference on Human Factors in Computing Systems*. New York, NY, USA: Association for Computing Machinery, 2021.
99. J. D. Eveland, Anita Blanchard, William Brown, and Jennifer Mattocks. „The role of "help networks" in facilitating use of CSCW tools". In: *CSCW '94: Proceedings of the*

1994 ACM conference on Computer supported cooperative work. New York, NY, USA: Association for Computing Machinery, Oct. 1994, pp. 265–274.
100. Olujoke A. Fakoya, Noleen K. McCorry, and Michael Donnelly. „Loneliness and social isolation interventions for older adults: a scoping review of reviews". In: *BMC Public Health* 20.1 (Feb. 2020), p. 129. eprint: 32054474.
101. Juan Fasola and Maja J Mataric. „Robot exercise instructor: A socially assistive robot system to monitor and encourage physical exercise for the elderly". In: *19th International Symposium in Robot and Human Interactive Communication*. IEEE. 2010, pp. 416–421.
102. Micha Fedrowitz and Sabine Matzke. „Das gemeinschaftliche Wohnen für Ältere". In: *Informationen zur Raumentwicklung* 2 (2013), pp. 177–187.
103. David Feil-Seifer and Maja J Mataric. „Defining socially assistive robotics". In: *9th International Conference on Rehabilitation Robotics, 2005. ICORR 2005*. IEEE. 2005, pp. 465–468.
104. Ronit Feingold Polak and Shelly Levy Tzedek. „Social Robot for Rehabilitation: Expert Clinicians and Post-Stroke Patients' Evaluation Following a Long-Term Intervention". In: HRI '20. Cambridge, United Kingdom: Association for Computing Machinery, 2020, pp. 151–160.
105. Anja Feldmann, Oliver Gasser, Franziska Lichtblau, et al. „The Lockdown Effect: Implications of the COVID-19 Pandemic on Internet Traffic". In: *Proceedings of the ACM Internet Measurement Conference*. IMC '20. Virtual Event, USA: Association for Computing Machinery, 2020, pp. 1–18.
106. Kenneth F Ferraro. „Self-ratings of health among the old and the old-old". In: *Journal of Health and Social Behavior* (1980), pp. 377–383.
107. Laura Fiorini, Kasia Tabeau, Grazia D'Onofrio, et al. „Co-creation of an assistive robot for independent living: lessons learned on robot design". In: *Int. J. Interact. Des. Manuf.* 14.2 (2020), pp. 491–502.
108. Shira H. Fischer, Daniel David, Bradley H. Crotty, Meghan Dierks, and Charles Safran. „Acceptance and Use of Health Information Technology By Community-Dwelling Elders". In: *Int. J. Med. Inf.* 83.9 (Sept. 2014), p. 624.
109. Luciano Floridi and Jeff W Sanders. „On the morality of artificial agents". In: *Minds and machines* 14.3 (2004), pp. 349–379.
110. Milton Friedman. *Capitalism and Freedom (40th Anniversary Edition)*. Chicago, IL, USA: University of Chicago Press, 2002.
111. Shinya Fujie, Yoichi Matsuyama, Hikaru Taniyama, and Tetsunori Kobayashi. „Conversation robot participating in and activating a group communication". In: *Tenth Annual Conference of the International Speech Communication Association*. 2009.
112. Gaaitzen J. de Vries, Elisabetta Gentile, Sébastien Miroudot, and Konstantin M. Wacker. „The rise of robots and the fall of routine jobs". In: *Labour Economics* 66 (2020), p. 101885.
113. Ann Gallagher, Dagfinn Nåden, and Dag Karterud. *Robots in elder care: Some ethical questions*. 2016.
114. Vera Gallistl, Alexander Seifert, and Franz Kolland. „COVID-19 as a "Digital Push?" Research Experiences From Long-Term Care and Recommendations for the Post-pandemic Era". In: *Front Public Health* 9 (May 2021).

115. Herbert J Gans. „Participant observation in the era of “ethnography”“. In: *Journal of contemporary ethnography* 28.5 (1999), pp. 540–548.
116. Bill Gates. „A robot in every home“. In: *Scientific American* 296.1 (2007), pp. 58–65.
117. GENESIS-Online. *Bevölkerung: Altersjahre, Geschlecht*. 2020.
118. Silvia Gerisch, Kornelius Knapp, and Karin Töpsch. *Demografiefeste Personalpolitik in der Altenpflege: Handlungsbedarf erfassen*. Vol. 39. W. Bertelsmann Verlag, 2009. Chap. Herausforderungen für die Pflege, pp. 9–23.
119. Elizabeth Gibney. „Google AI algorithm masters ancient game of Go“. In: *Nature* 529.7587 (Jan. 2016), pp. 445–446.
120. Jennifer Goetz, Sara Kiesler, and Aaron Powers. „Matching robot appearance and behavior to tasks to improve human-robot cooperation“. In: *The 12th IEEE International Workshop on Robot and Human Interactive Communication, 2003. Proceedings. ROMAN 2003*. Ieee. 2003, pp. 55–60.
121. Cristopher Gómez, Matìas Mattamala, Tim Resink, and Javier Ruiz-del-Solar. „Visual slam-based localization and navigation for service robots: The pepper case“. In: *Robot World Cup*. Springer. 2018, pp. 32–44.
122. Rachael Gooberman-Hill and Shah Ebrahim. „Informal care at times of change in health and mobility: a qualitative study“. In: *Age and Ageing* 35.3 (2006), pp. 261–266.
123. Anne Goodhead and Janet McDonald. „Informal caregivers literature review“. In: *A report prepared for the National Health Committee Health Services Research Centre: Victoria University of Wellington* (2007).
124. Stefan Görres, Kathrin Seibert, and Susanne Stiefler. „Perspektiven zum pflegerischen Versorgungsmix“. In: 2016, pp. 3–18.
125. Maartje de Graaf, Somaya Ben Allouch, and Jan van Dijk. „Why Do They Refuse to Use My Robot? Reasons for Non-Use Derived from a Long-Term Home Study“. In: *Proceedings of the 2017 ACM/IEEE International Conference on Human-Robot Interaction*. HRI ’17. New York, NY, USA: ACM, 2017, pp. 224–233.
126. Maartje M. A. de Graaf, Somaya Ben Allouch, and Tineke Klamer. „Sharing a life with Harvey: Exploring the acceptance of and relationship-building with a social robot“. In: *Computers in Human Behavior* 43 (Feb. 2015), pp. 1–14.
127. Birgit Graf, Matthias Hans, and Rolf D Schraft. „Care-O-bot II—Development of a next generation robotic home assistant“. In: *Autonomous robots* 16.2 (2004), pp. 193–205.
128. Gregor Wolbring and Sophya Yumakulov. „Social Robots: Views of Staff of a Disability Service Organization“. In: *International Journal of Social Robotics* 6.3 (2014), pp. 457–468.
129. Peter Griffiths, Chiara Dall’Ora, Michael Simon, et al. „Nurses’ shift length and overtime working in 12 European countries: the association with perceived quality of care and patient safety“. In: *Medical care* 52.11 (2014), p. 975.
130. Kristina Groth and Jeremiah Scholl. „Coordination in highly-specialized care networks“. In: *CSCW ’13: Proceedings of the 2013 conference on Computer supported cooperative work companion*. New York, NY, USA: Association for Computing Machinery, Feb. 2013, pp. 143–148.
131. Gunnar Stevens and Volkmar Pipek. *Making Use: Understanding, Studying and supporting appropriation*. Oxford University Press, 2018.

132. Loni Hagen, Stephen Neely, Ryan Scharf, and Thomas E. Keller. „Social Media Use for Crisis and Emergency Risk Communications During the Zika Health Crisis". In: *Digit. Gov.: Res. Pract.* 1.2 (Apr. 2020).
133. Marc Hanheide, Denise Hebesberger, and Tomáš Krajn\'ik. „The When, Where, and How: An Adaptive Robotic Info-Terminal for Care Home Residents". In: *Proceedings of the 2017 ACM/IEEE International Conference on Human-Robot Interaction*. HRI '17. New York, NY, USA: ACM, 2017, pp. 341–349.
134. Dave Harley, Julie Morgan, and Hannah Frith. „Growing Older". In: *Cyberpsychology as Everyday Digital Experience across the Lifespan*. London, England, UK: Palgrave Macmillan, May 2018, pp. 175–198.
135. Jean D. Hallewell Haslwanter, Katja Neureiter, and Markus Garschall. „User-centered design in AAL". In: *Univ. Access Inf. Soc.* 19.1 (Mar. 2020), pp. 57–67.
136. Dirk Heckmann and Anne Paschke. „Datenschutzrechtliche Aspekte von Big Data-Analysen im Gesundheitswesen". In: *Schriftenreihe der Stiftung Datenschutz, Daten-Debatten Band 2 Big Data und E-Health*. Erich Schmidt Verlag, 2017.
137. Christoph Heible. Springer-Verlag, 2015, pp. 135–319.
138. Thomas Heinze. „How to sponsor ground-breaking research: a comparison of funding schemes". In: *Science and public policy* 35.5 (2008), pp. 302–318.
139. Helinä Melkas, Lea Hennala, Satu Pekkarinen, and Ville Kyrki. „Impacts of robot implementation on care personnel and clients in elderly-care institutions". In: *International Journal of Medical Informatics* 134 (2020), p. 104041.
140. Melina Helm, Felix Carros, Johannes Schädler, and Volker Wulf. „Zoomorphic Robots and People with Disabilities". In: *Proceedings of Mensch und Computer 2022*. 2022, pp. 431–436.
141. Anna Henschel, Guy Laban, and Emily S Cross. „What makes a robot social? a review of social robots from science fiction to a home or hospital near you". In: *Current Robotics Reports* 2.1 (2021), pp. 9–19.
142. Abdelfetah Hentout, Mustapha Aouache, Abderraouf Maoudj, and Isma Akli. „Human–robot interaction in industrial collaborative robotics: a literature review of the decade 2008–2017". In: *Advanced Robotics* 33.15-16 (2019), pp. 764–799.
143. Patrick Hock, Chika Oshima, and Koichi Nakayama. „CATARO: a robot that tells caregivers a patient's current non-critical condition indirectly". In: *GECCO '18: Proceedings of the Genetic and Evolutionary Computation Conference Companion*. New York, NY, USA: Association for Computing Machinery, 2018, pp. 1841–1844.
144. Neil G Hockstein, CG Gourin, RA Faust, and David J Terris. „A history of robots: from science fiction to surgical robotics". In: *Journal of robotic surgery* 1.2 (2007), pp. 113–118.
145. Ulrike Höhmann, Manuela Lautenschläger, and Laura Schwarz. „Belastungen im Pflegeberuf: Bedingungsfaktoren, Folgen und Desiderate". In: *Pflege-Report* (2016), pp. 73–89.
146. Eva Hornecker, Andreas Bischof, Philipp Graf, Lena Franzkowiak, and Norbert Krüger. „The Interactive Enactment of Care Technologies and Its Implications for Human-Robot-Interaction in Care". In: *Proceedings of the 11th Nordic Conference on Human-Computer Interaction: Shaping Experiences, Shaping Society*. NordiCHI '20. New York, NY, USA: Association for Computing Machinery, 2020.

147. Dominik Hornung, Claudia Müller, Alexander Boden, and Martin Stein. „Autonomy Support for Elderly People through Everyday Life Gadgets". In: *GROUP '16: Proceedings of the 19th International Conference on Supporting Group Work*. New York, NY, USA: Association for Computing Machinery, Nov. 2016, pp. 421–424.
148. Ayanna Howard. „Are we trusting AI too much? Examining human-robot interactions in the real world". In: *2020 15th ACM/IEEE International Conference on Human-Robot Interaction (HRI)*. IEEE. 2020, pp. 1–1.
149. John A Hughes, Dave Randall, and Dan Shapiro. „From ethnographic record to system design". In: *Computer Supported Cooperative Work (CSCW)* 1.3 (1992), pp. 123–141.
150. Kimberly A Hughes and Rose M Reynolds. „Evolutionary and mechanistic theories of aging". In: *Annual review of entomology* 50.1 (2005), pp. 421–445.
151. Julie Huschilt and Laurie Clune. „The use of socially assistive robots for dementia care". In: *J. Gerontol. Nurs.* 38.10 (2012), pp. 15–19.
152. Christos Iavazzo, Xanthi-Ekaterini D Gkegke, Paraskevi-Evangelia Iavazzo, and Ioannis D Gkegkes. „Evolution of robots throughout history from Hephaestus to Da Vinci Robot". In: *AMHA-Acta medico-historica Adriatica* 12.2 (2014), pp. 247–258.
153. Sumio Ishii. „Meal-assistance robot "my spoon"". In: *Journal of the Robotics Society of Japan* 21.4 (2003), pp. 378–381.
154. Hirokazu Ito, M. Miyagawa, Y. Kuwamura, et al. *Professional Nurses' Attitudes towards the Introduction of Humanoid Nursing Robots (HNRs) in Health Care Settings*. 2015.
155. Klaus Jacobs, Antje Schwinger, Jürgen Klauber, Stefan Greß, and Adelheid Kuhlmey. *Pflege-Report 2016: Schwerpunkt: Die Pflegenden im Fokus*. Schattauer Verlag, 2016.
156. Kathrin Janowski, Hannes Ritschel, Birgit Lugrin, and Elisabeth André. „Sozial interagierende Roboter in der Pflege". In: *Pflegeroboter*. Springer Gabler, Wiesbaden, 2018, pp. 63–87.
157. Jérémy Barande, Wikipedia. *Nao - Vivatech 2019*. [Online; accessed June 17, 2022]. 2019.
158. Carolyn Johnston. „Ethical design and use of robotic care of the elderly". In: *Journal of Bioethical Inquiry* 19.1 (2022), pp. 11–14.
159. Brigitte Jordan. „Ethnographic workplace studies and CSCW". In: *Human Factors in Information Technology*. Vol. 12. Elsevier, 1996, pp. 17–42.
160. Joseph Andrew Pepito and Rozzano Locsin. „Can nurses remain relevant in a technologically advanced future?" In: *International Journal of Nursing Sciences* 6.1 (2019), pp. 106–110.
161. Suhas Govind Joshi and Tone Bratteteig. „Designing for prolonged mastery. On involving old people in participatory design". In: *Scandinavian Journal of Information Systems* 28.1 (2016), p. 1.
162. Reza Kachouie, Sima Sedighadeli, Rajiv Khosla, and Mei-Tai Chu. „Socially assistive robots in elderly care: a mixed-method systematic literature review". In: *International Journal of Human-Computer Interaction* 30.5 (2014), pp. 369–393.
163. Masao Kanamori, Mizue Suzuki, and Misao Tanaka. „Maintenance and improvement of quality of life among elderly patients using a pet-type robot". In: *Nihon Ronen Igakkai zasshi. Japanese journal of geriatrics* 39.2 (2002), pp. 214–218.
164. Junko Kanero, Vasfiye Geçkin, Cansu Oranç, et al. „Social robots for early language learning: Current evidence and future directions". In: *Child Development Perspectives* 12.3 (2018), pp. 146–151.

165. Helena Karasti. „Increasing sensitivity towards everyday work practice in system design“. In: (2001).
166. Azmina Karimi and Carman Neustaedter. „From high connectivity to social isolation: communication practices of older adults in the digital age“. In: *CSCW '12: Proceedings of the ACM 2012 conference on Computer Supported Cooperative Work Companion*. New York, NY, USA: Association for Computing Machinery, Feb. 2012, pp. 127–130.
167. Kathleen England and Natasha Azzopardi-Muscat. „Demographic trends and public health in Europe“. In: *European Journal of Public Health* 27.suppl_4 (2017), pp. 9–13.
168. Christian Katzenmeier. „Big data, E-health, M-health, KI und Robotik in der Medizin“. In: *Medizinrecht* 37.4 (2019), pp. 259–271.
169. Christoph Kehl. „Robotik und assistive Neurotechnologien in der Pflege–gesellschaftliche Herausforderungen“. In: *Vertiefung des Projekts 'Mensch-Maschine-Entgrenzung'. Bericht des Ausschusses für cätzung (18. Ausschuss) gemäß §56a der Geschäftsordnung. Berlin Deutscher Bundestag (Drucksache 19/2790)* (2018).
170. Finn Kensing and Joan Greenbaum. „Heritage: Having a say“. In: *Routledge international handbook of participatory design*. Routledge, 2012, pp. 41–56.
171. Rajiv Khosla, Khanh Nguyen, and Mei-Tai Chu. „Human robot engagement and acceptability in residential aged care“. In: *International Journal of Human–Computer Interaction* 33.6 (2017), pp. 510–522.
172. C. D. Kidd, W. Taggart, and S. Turkle. „A sociable robot to encourage social interaction among the elderly“. In: *Proceedings 2006 IEEE International Conference on Robotics and Automation, 2006. ICRA 2006*. 2006, pp. 3972–3976.
173. Seongseop Sam Kim, Jungkeun Kim, Frank Badu-Baiden, Marilyn Giroux, and Youngjoon Choi. „Preference for robot service or human service in hotels? Impacts of the COVID-19 pandemic“. In: *International Journal of Hospitality Management* 93 (2021), p. 102795.
174. Tess Knight and Lina A Ricciardelli. „Successful aging: Perceptions of adults aged between 70 and 101 years“. In: *The International Journal of Aging and Human Development* 56.3 (2003), pp. 223–245.
175. Yaacov Koren, Ronit Feingold Polak, and Shelly Levy-Tzedek. „Extended Interviews with Stroke Patients Over a Long-Term Rehabilitation Using Human–Robot or Human–Computer Interactions“. In: *International Journal of Social Robotics* 14.8 (2022), pp. 1893–1911.
176. Kostas Koutsoumpas, Vasilis Kostogiannis, Konstantinos Zis, and Athanasios Kotsakis. „Integrated Care for the Elderly: A Review of the Literature“. In: *Healthcare* 9.3 (Mar. 2020), p. 244.
177. Naoki Koyama, Kazuaki Tanaka, Kohei Ogawa, and Hiroshi Ishiguro. „Emotional or Social? How to Enhance Human-Robot Social Bonding“. In: *HAI '17: Proceedings of the 5th International Conference on Human Agent Interaction*. New York, NY, USA: Association for Computing Machinery, 2017, pp. 203–211.
178. R Hari Krishnan and S Pugazhenthi. „Mobility assistive devices and self-transfer robotic systems for elderly, a review“. In: *Intelligent Service Robotics* 7.1 (2014), pp. 37–49.
179. Udo Kuckartz. *Mixed methods: methodologie, Forschungsdesigns und Analyseverfahren*. Springer-Verlag, 2014.
180. Jürgen Kühling. „Datenschutz im Gesundheitswesen“. In: *Medizinrecht* 37.8 (2019), pp. 611–622.

181. Robert V Labaree. „The risk of 'going observationalist': negotiating the hidden dilemmas of being an insider participant observer". In: *Qualitative research* 2.1 (2002), pp. 97–122.
182. Janne Lahtiranta. „Mediator–enabler for successful digital health care". In: *Finnish Journal of eHealth and eWelfare* 9.4 (2017), pp. 284–298.
183. Linne Marie Lauesen. „Secondary stakeholders". In: *Encyclopedia of Corporate Social Responsibility*. Springer Science+ Business Media, 2013, pp. 2114–2119.
184. Jonathan Lazar. „Managing digital accessibility at universities during the COVID-19 pandemic". In: *Univ. Access Inf. Soc.* (Mar. 2021), pp. 1–17.
185. Tiago Leal Ghezzi and Oly Campos Corleta. „30 years of robotic surgery". In: *World journal of surgery* 40.10 (2016), pp. 2550–2557.
186. Hee Rin Lee, Selma Šabanović, Wan-Ling Chang, et al. „Steps Toward Participatory Design of Social Robots: Mutual Learning with Older Adults with Depression". In: *Proceedings of the 2017 ACM/IEEE International Conference on Human-Robot Interaction*. HRI '17. New York, NY, USA: ACM, 2017, pp. 244–253.
187. Jasmin Lehmann, Felix Carros, David Unbehaun, Rainer Wieching, and Jens Lüssem. „Einsatzfelder der sozialen Robotik in der Pflege". In: *N. Krämer, C. Stoffers, & C. Heitmann (Hsrg.), Digitale Transformation im Krankenhaus. Thesen, Potenziale, Anwendungen* (2019), pp. 88–113.
188. Iolanda Leite, Carlos Martinho, and Ana Paiva. „Social Robots for Long-Term Interaction: A Survey". In: *Int. J. Social Rob.* 5.2 (2013), pp. 291–308.
189. Przemysław Lewicki, Jacek Tochowicz, and Jeroen van Genuchten. „Are Robots Taking Our Jobs? A RoboPlatform at a Bank". In: *IEEE Software* 36.3 (2019), pp. 101–104.
190. Myriam Lewkowicz. „Strategic thinking for sustainable practice-centered computing". In: *Future-Proofing: Making Practice-based IT Design Sustainable*. Oxford University Press, 2022.
191. Henry Lieberman, Fabio Paternò, Markus Klann, and Volker Wulf. „End-user development: An emerging paradigm". In: *End user development*. Springer, 2006, pp. 1–8.
192. Gesa Lindemann. „Social interaction with robots: three questions". In: *AI & society* 31.4 (2016), pp. 573–575.
193. Stephen Lindsay, Daniel Jackson, Guy Schofield, and Patrick Olivier. „Engaging Older People Using Participatory Design". In: *Proceedings of the SIGCHI Conference on Human Factors in Computing Systems*. New York, NY, USA: Association for Computing Machinery, 2012, pp. 1199–1208.
194. Wing-Yue Geoffrey Louie, Derek McColl, and Goldie Nejat. „Acceptance and attitudes toward a human-like socially assistive robot by older adults". In: *Assistive Technology* 26.3 (2014), pp. 140–150.
195. Andra Lutu, Diego Perino, Marcelo Bagnulo, Enrique Frias-Martinez, and Javad Khangosstar. „A Characterization of the COVID-19 Pandemic Impact on a Mobile Network Operator Traffic". In: *Proceedings of the ACM Internet Measurement Conference*. IMC '20. Virtual Event, USA: Association for Computing Machinery, 2020, pp. 19–33.
196. Maartje M.A. de Graaf, Somaya Ben Allouch, and Tineke Klamer. „Sharing a life with Harvey: Exploring the acceptance of and relationship-building with a social robot". In: *Computers in Human Behavior* 43 (2015), pp. 1–14.
197. Arne Maibaum, Andreas Bischof, Jannis Hergesell, and Benjamin Lipp. „A critique of robotics in health care". In: *AI & SOCIETY* 37.2 (2022), pp. 467–477.

198. José Ginés Giménez Manuel, Juan Carlos Augusto, and Jill Stewart. „AnAbEL: towards empowering people living with dementia in ambient assisted living". In: *Univ. Access Inf. Soc.* (Nov. 2020), pp. 1–20.
199. Arne Manzeschke. „MEESTAR: ein Modell angewandter Ethik im Bereich assistiver Technologien". In: *Technisierung des Alters–Beitrag zu einem guten Leben* (2015), pp. 263–283.
200. Arne Manzeschke, Karsten Weber, Elisabeth Rother, and Heiner Fangerau. *Ethische Fragen im Bereich Altersgerechter Assistenzsysteme*. VDI/VDE, 2013.
201. H. R. Marston, Loredana Ivan, M. Fernández-Ardévol, et al. „COVID-19: Technology, Social Connections, Loneliness, and Leisure Activities: An International Study Protocol". In: *undefined* (2020).
202. Angela Mastrianni, Leah Kulp, and Aleksandra Sarcevic. „Transitioning to Remote User-Centered Design Activities in the Emergency Medical Field During a Pandemic". In: *Extended Abstracts of the 2021 CHI Conference on Human Factors in Computing Systems*. CHI EA '21. Yokohama, Japan: Association for Computing Machinery, 2021.
203. David Maulsby, Saul Greenberg, and Richard Mander. „Prototyping an intelligent agent through Wizard of Oz". In: *Proceedings of the INTERACT'93 and CHI'93 conference on Human factors in computing systems*. 1993, pp. 277–284.
204. Victoria McGeer. „Mind-making practices: The social infrastructure of self-knowing agency and responsibility". In: *Philosophical Explorations* 18.2 (2015), pp. 259–281.
205. Victoria McGeer. „The regulative dimension of folk psychology". In: *Folk psychology re-assessed*. Springer, 2007, pp. 137–156.
206. Johanna Meurer, Claudia Müller, Carla Simone, Ina Wagner, and Volker Wulf. „Designing for sustainability: Key issues of ICT projects for ageing at home". In: *Computer Supported Cooperative Work (CSCW)* 27.3 (2018), pp. 495–537.
207. Christine Milligan, Celia Roberts, and Maggie Mort. „Telecare and older people: Who cares where?" In: *Social science & medicine (1982)* 72.3 (2011), pp. 347–354.
208. Catrin Misselhorn. „Artificial systems with moral capacities? A research design and its implementation in a geriatric care system". In: *Artificial Intelligence* 278 (2020), p. 103179.
209. Catrin Misselhorn. „Digitale Rechtssubjekte, Handlungsfähigkeit und Verantwortung aus philosophischer Sicht". In: *Verfassungsblog: On Matters Constitutional* (2019).
210. Catrin Misselhorn. *Grundfragen der Maschinenethik: Reclams Universal-Bibliothek*. Reclam Verlag, 2018.
211. Catrin Misselhorn. „Robots as moral agents". In: *Roboethics, proceedings of the annual conference on ethics of the German association for social science research on Japan*. Iudicum München. 2013, pp. 30–42.
212. Marc Mitchell and Lena Kan. „Digital technology and the future of health systems". In: *Health Systems & Reform* 5.2 (2019), pp. 113–120.
213. Tracy L. Mitzner, Charles C. Kemp, Wendy Rogers, and Lorenza Tiberio. „Investigating Healthcare Providers' Acceptance of Personal Robots for Assisting with Daily Caregiving Tasks". In: *CHI '13 Extended Abstracts on Human Factors in Computing Systems*. CHI EA '13. New York, NY, USA: Association for Computing Machinery, 2013, pp. 499–504.
214. Sanika Moharana, Alejandro E. Panduro, Hee Rin Lee, and Laurel D. Riek. „Robots for joy, robots for sorrow: community based robot design for dementia caregivers". In: *HRI*

'19: Proceedings of the 14th ACM/IEEE International Conference on Human-Robot Interaction. IEEE Press, 2019, pp. 458–467.

215. Masahiro Mori, Karl F. MacDorman, and Norri Kageki. „The Uncanny Valley [From the Field]". In: *IEEE Robotics & Automation Magazine* 19.2 (2012), pp. 98–100.
216. Claudia Müller, Cornelius Neufeldt, David Randall, and Volker Wulf. „ICT-Development in Residential Care Settings: Sensitizing Design to the Life Circumstances of the Residents of a Care Home". In: CHI '12. Austin, Texas, USA: Association for Computing Machinery, 2012, pp. 2639–2648.
217. Claudia Müller, Marén Schorch, and Rainer Wieching. „PraxLabs as a Setting for Participatory Technology Research and Design in the Field of HRI and Demography". In: *Proceedings of the workshop „socially assistive robots for the aging population: are we trapped in stereotypes*. 2014.
218. Maurice Mulvenna, Suzanne Martin, Donal McDade, et al. „TRAIL Living Labs Survey 2011: A survey of the ENOLL living labs". In: (2011).
219. Sara Nadin and Catherine Cassell. „The use of a research diary as a tool for reflexive practice: Some reflections from management research". In: *Qualitative Research in Accounting & Management* (2006).
220. Junya Nakanishi, Itaru Kuramoto, Jun Baba, et al. „Continuous Hospitality with Social Robots at a hotel". In: *SN Applied Sciences* 2.3 (2020), pp. 1–13.
221. Toyoshiro Nakashima, Goro Fukutome, and Naohiro Ishii. „Healing effects of pet robots at an elderly-care facility". In: *2010 IEEE/ACIS 9th International Conference on Computer and Information Science*. IEEE. 2010, pp. 407–412.
222. Engineering National Academies of Sciences, Of Behavioral, Social Sciences, et al. „Social Isolation and Loneliness in Older Adults: Opportunities for the Health Care System". In: *PubMed* (2021). eprint: 32510896.
223. United Nations. „World population ageing 2009: United Nations, Dept. of Economic and Social Affairs". In: *Population Division* (2010).
224. Goldie Nejat, Yiyuan Sun, and Mary Nies. „Assistive Robots in Health Care Settings". In: *Home Health Care Management & Practice* 21.3 (2008), pp. 177–187.
225. Kathryn E Newcomer, Harry P Hatry, and Joseph S Wholey. „Conducting semi-structured interviews". In: *Handbook of practical program evaluation* 492 (2015), p. 492.
226. Chi-hung Ng. „Motivation among older adults in learning computing technologies: A grounded model". In: *Educational Gerontology* 34.1 (2007), pp. 1–14.
227. Linh Cuong Nguyen. „The impact of humanoid robots on Australian public libraries". In: *Journal of the Australian Library and Information Association* 69.2 (2020), pp. 130–148.
228. Matthias Nieuwenhuisen, Jörg Stückler, and Sven Behnke. „Intuitive multimodal interaction for service robots". In: *HRI '10: Proceedings of the 5th ACM/IEEE international conference on Human-robot interaction*. IEEE Press, Mar. 2010, pp. 177–178.
229. Jay F Nunamaker Jr, Robert O Briggs, Douglas C Derrick, and Gerhard Schwabe. „The last research mile: Achieving both rigor and relevance in information systems research". In: *Journal of management information systems* 32.3 (2015), pp. 10–47.
230. Kristen Nygaard and Olav Terje Bergo. „The Trade Unions-New users of research". In: *Personnel review* (1975).

231. Sven Nyholm and Lily Eva Frank. „It loves me, it loves me not: is it morally problematic to design sex robots that appear to love their owners?" In: *Techné : Research in Philosophy and Technology* (2019).
232. Rory O'Brien. „An overview of the methodological approach of action research". In: (1998).
233. Jim Oeppen and James W. Vaupel. „Broken Limits to Life Expectancy". In: *Science* 296.5570 (2002), pp. 1029–1031. eprint: https://www.science.org/doi/pdf/10.1126/science.1069675.
234. Federal Statistical Office. June 2019.
235. Corinna Ogonowski, Timo Jakobi, Jan Hess, et al. „PRAXLABS: A Sustainable Framework for User-Centered Information and Communication Technology Development-Cultivating Research Experiences from Living Labs in the Home". In: *Wulf, Pipek et al.(Eds.): Socio-Informatics: A Practice-Based Perspective on the Design and Use of IT Artifacts*. Oxford University Press, 2018, pp. 319–360.
236. Corinna Ogonowski, Benedikt Ley, Jan Hess, Lin Wan, and Volker Wulf. „Designing for the Living Room: Long-Term User Involvement in a Living Lab". In: CHI '13. Paris, France: Association for Computing Machinery, 2013, pp. 1539–1548.
237. Amalya L Oliver. „Holistic ecosystems for enhancing innovative collaborations in university–industry consortia". In: *The Journal of Technology Transfer* (2022), pp. 1–17.
238. Linda Onnasch, Xenia Maier, and Thomas Jürgensohn. *Mensch-Roboter-Interaktion-Eine Taxonomie für alle Anwendungsfälle*. Bundesanstalt für Arbeitsschutz und Arbeitsmedizin Dortmund, 2016.
239. K. Oswald. *Citizen Science auf Instagram*. n. d.
240. Oxford English Dictionary. *„robot, n.2"*. Accessed: 2022-05-27.
241. Leysia Ann Palen. „Groupware Adoption and Adaptation". In: *CHI '97 Extended Abstracts on Human Factors in Computing Systems*. CHI EA '97. New York, NY, USA: Association for Computing Machinery, 1997, pp. 67–68.
242. Yadong Pan, Haruka Okada, Toshiaki Uchiyama, and Kenji Suzuki. „Direct and indirect social robot interactions in a hotel public space". In: *2013 IEEE International Conference on Robotics and Biomimetics (ROBIO)*. IEEE. 2013, pp. 1881–1886.
243. Amit Kumar Pandey and Rodolphe Gelin. „A mass-produced sociable humanoid robot: Pepper: The first machine of its kind". In: *IEEE Robotics & Automation Magazine* 25.3 (2018), pp. 40–48.
244. Jennifer A Parks. „Lifting the burden of Women's care work: should robots replace the "human touch"?" In: *Hypatia* 25.1 (2010), pp. 100–120.
245. European Parliament. „REPORT with recommendations to the Commission on Civil Law Rules on Robotics (2015/2103(INL)". In: (2017).
246. Sakari Pieska, Mika Luimula, Juhana Jauhiainen, and Van Spiz. „Social service robots in wellness and restaurant applications". In: *Journal of Communication and Computer* 10.1 (2013), pp. 116–123.
247. Maribel Pino, Me\ifmmode\mboxç\elseç\fiois Boulay, and Anne-Sophie Rigaud. „"Are we ready for robots that care for us?" Attitudes and opinions of older adults toward socially assistive robots". In: *Front. Aging Neurosci.* 7 (2015), p. 141.
248. Volkmar Pipek and Volker Wulf. „Infrastructuring: Toward an integrated perspective on the design and use of information technology". In: *Journal of the Association for Information Systems* 10.5 (2009), p. 1.

249. Martha E Pollack, Laura Brown, Dirk Colbry, et al. „Pearl: A mobile robotic assistant for the elderly“. In: *AAAI workshop on automation as eldercare*. AAAI Canada. 2002.
250. Akanksha Prakash, Jenay M Beer, Travis Deyle, et al. „Older adults' medication management in the home: How can robots help?“ In: *2013 8th ACM/IEEE International Conference on Human-Robot Interaction (HRI)*. IEEE. 2013, pp. 283–290.
251. Rob Procter, Trisha Greenhalgh, Joe Wherton, et al. „The Day-to-Day Co-Production of Ageing in Place“. In: *Comput. Supported Coop. Work* 23.3 (June 2014), pp. 245–267.
252. Lihui Pu, Wendy Moyle, Cindy Jones, and Michael Todorovic. „The effectiveness of social robots for older adults: a systematic review and meta-analysis of randomized controlled studies“. In: *The Gerontologist* 59.1 (2019), e37–e51.
253. David Randall, Richard Harper, and Mark Rouncefield. *Fieldwork for design: theory and practice*. Springer Science & Business Media, 2007.
254. Penny Rhodes and Sandra Shaw. „Informal care and terminal illness“. In: *Health & social care in the community* 7.1 (1999), pp. 39–50.
255. Nina Riether, Frank Hegel, Britta Wrede, and Gernot Horstmann. „Social facilitation with social robots?“ In: *2012 7th ACM/IEEE International Conference on Human-Robot Interaction (HRI)*. IEEE. 2012, pp. 41–47.
256. Giuseppe Riva and Eleonora Riva. „CARESSES: The world's first culturally sensitive robots for elderly care“. In: (2019).
257. Rob McCarney, James Warner, Steve Iliffe, et al. „The Hawthorne Effect: a randomised, controlled trial“. In: *BMC Medical Research Methodology* 7.1 (2007).
258. Toni Robertson and Jesper Simonsen. „Participatory Design: an introduction“. In: *Routledge international handbook of participatory design*. Routledge, 2012, pp. 1–17.
259. Toni Robertson and Ina Wagner. „Ethics: Engagement, representation and politics-in-action“. In: *Routledge international handbook of participatory design*. Routledge, 2012, pp. 84–105.
260. Paul Robinette, Ayanna Howard, and Alan R Wagner. „Conceptualizing overtrust in robots: why do people trust a robot that previously failed?“ In: *Autonomy and Artificial Intelligence: A Threat or Savior?* Springer, 2017, pp. 129–155.
261. Paul Robinette, Wenchen Li, Robert Allen, Ayanna M Howard, and Alan R Wagner. „Overtrust of robots in emergency evacuation scenarios“. In: *2016 11th ACM/IEEE international conference on human-robot interaction (HRI)*. IEEE. 2016, pp. 101–108.
262. Laura Robinson, Jeremy Schulz, Aneka Khilnani, et al. „Digital inequalities in time of pandemic: COVID-19 exposure risk profiles and new forms of vulnerability“. In: *First Monday* 25.10 (July 2020).
263. Lisa Charlotte Rost. *A rusting fountain of youth in Germany*. Nov. 2017.
264. Leonel Rozo, Sylvain Calinon, Darwin G Caldwell, Pablo Jimenez, and Carme Torras. „Learning physical collaborative robot behaviors from human demonstrations“. In: *IEEE Transactions on Robotics* 32.3 (2016), pp. 513–527.
265. Selma Šabanović, Casey C Bennett, Wan-Ling Chang, and Lesa Huber. „PARO robot affects diverse interaction modalities in group sensory therapy for older adults with dementia“. In: *2013 IEEE 13th international conference on rehabilitation robotics (ICORR)*. IEEE. 2013, pp. 1–6.
266. Selma Šabanović and Wan-Ling Chang. „Socializing robots: constructing robotic sociality in the design and use of the assistive robot PARO“. In: *AI and Society* 31.4 (2016), pp. 537–551.

267. Selma Šabanović, Wan-Ling Chang, Casey C Bennett, Jennifer A Piatt, and David Hakken. „A robot of my own: participatory design of socially assistive robots for independently living older adults diagnosed with depression". In: *International conference on human aspects of it for the aged population*. Springer. 2015, pp. 104–114.
268. Alessandra Maria Sabelli, Takayuki Kanda, and Norihiro Hagita. „A conversational robot in an elderly care center: an ethnographic study". In: *HRI '11: Proceedings of the 6th international conference on Human-robot interaction*. New York, NY, USA: Association for Computing Machinery, 2011, pp. 37–44.
269. Åke Sandberg. *Computers dividing man and work: Recent Scandinavian research on planning and computers from a trade union perspective*. Arbetslivcentrum, 1979.
270. Tracy L Sanders, Tarita Wixon, K Elizabeth Schafer, Jessie YC Chen, and Peter A Hancock. „The influence of modality and transparency on trust in human-robot interaction". In: *2014 IEEE International Inter-Disciplinary Conference on Cognitive Methods in Situation Awareness and Decision Support (CogSIMA)*. IEEE. 2014, pp. 156–159.
271. Eduardo Benitez Sandoval, Omar Mubin, and Mohammad Obaid. „Human robot interaction and fiction: A contradiction". In: *International Conference on Social Robotics*. Springer. 2014, pp. 54–63.
272. Kjeld Schmidt. „Practice and technology: on the conceptual foundations of practice-centered computing". In: *Socio-informatics: A Practice-based Perspective on the Design and Use of IT Artifacts*. Oxford University Press, 2018, pp. 47–103.
273. Kjeld Schmidt. „Taking CSCW Seriously: Supporting Articulation Work (1992)". In: *Cooperative Work and Coordinative Practices*. London, England, UK: Springer, Sept. 2008, pp. 45–71.
274. Marén Schorch, Lin Wan, David William Randall, and Volker Wulf. „Designing for Those Who Are Overlooked: Insider Perspectives on Care Practices and Cooperative Work of Elderly Informal Caregivers". In: *Proceedings of the 19th ACM Conference on Computer-Supported Cooperative Work amp; Social Computing*. CSCW '16. San Francisco, California, USA: Association for Computing Machinery, 2016, pp. 787–799.
275. Douglas Schuler and Aki Namioka. *Participatory design: Principles and practices*. CRC Press, 1993.
276. Isabel Schwaninger, Felix Carros, Astrid Weiss, Volker Wulf, and Geraldine Fitzpatrick. „Video connecting families and social robots: from ideas to practices putting technology to work". In: *Universal Access in the Information Society* (2022).
277. Isabel Schwaninger, Geraldine Fitzpatrick, and Astrid Weiss. „Exploring Trust in Human-Agent Collaboration". In: *European Society for Socially Embedded Technologies (EUSSET)* (2019).
278. Isabel Schwaninger, Geraldine Fitzpatrick, and Astrid Weiss. „Exploring Trust in Human-Agent Collaboration". In: *European Society for Socially Embedded Technologies (EUSSET)* (2019).
279. Isabel Schwaninger, Christopher Frauenberger, and Geraldine Fitzpatrick. „Unpacking Forms of Relatedness around Older People and Telecare". In: *Companion Publication of the 2020 ACM Designing Interactive Systems Conference* (July 2020), pp. 163–169.
280. Isabel Schwaninger, Florian Güldenpfennig, Astrid Weiss, and Geraldine Fitzpatrick. „What Do You Mean by Trust? Establishing Shared Meaning in Interdisciplinary Design for Assistive Technology". In: *Int. J. Social Rob.* (2021), pp. 1–19.

281. Johanna Seibt. „Towards an ontology of simulated social interaction: varieties of the "As If" for robots and humans". In: *Sociality and normativity for robots*. Springer, 2017, pp. 11–39.
282. Alexander Seifert, Shelia R. Cotten, and Bo Xie. „A Double Burden of Exclusion? Digital and Social Exclusion of Older Adults in Times of COVID-19". In: *J. Gerontol. B Psychol. Sci. Soc. Sci.* 76.3 (Mar. 2021), e99–e103.
283. Kyung Hwa Seo and Jee Hye Lee. „The emergence of service robots at restaurants: Integrating trust, perceived risk, and satisfaction". In: *Sustainability* 13.8 (2021), p. 4431.
284. Kathia Serrano Velarde and Steffen Sigmund. „Drittmittel und die Rationalisierung der Wissenschaft. Eine Forschungsskizze". In: *Theorie als Beruf*. Springer, 2021, pp. 75–86.
285. Stuart Shepherd and Alois Buchstab. „Kuka robots on-site". In: *Robotic Fabrication in Architecture, Art and Design 2014*. Springer, 2014, pp. 373–380.
286. Takanori Shibata, Kazuya Ohkawa, and Kazuo Tanie. „Spontaneous behavior of robots for cooperation. Emotionally intelligent robot system". In: *Proceedings of IEEE International Conference on Robotics and Automation*. Vol. 3. IEEE. 1996, pp. 2426–2431.
287. Judith Sixsmith, A. Sixsmith, A. Malmgren Fänge, et al. „Healthy ageing and home: The perspectives of very old people in five European countries". In: *Soc. Sci. Med.* 106 (Apr. 2014), pp. 1–9.
288. Judith M. Smith. „Loneliness in older adults: an embodied experience". In: *J. Gerontol. Nurs.* 38.8 (Aug. 2012), pp. 45–53. eprint: 22800406.
289. Tim Smithers. „Autonomy in robots and other agents". In: *Brain and cognition* 34.1 (1997), pp. 88–106.
290. Victoria Song. „Useless Police Robot Fails to Call For Help When Needed". In: https://gizmodo.com (Aug. 2019).
291. Roger Andre Søraa, Pernille Søderholm Nyvoll, Karoline Blix Grønvik, and J. Artur Serrano. „Children's perceptions of social robots: a study of the robots Pepper, AV1 and Tessa at Norwegian research fairs". In: *AI & society* 36 (2021), pp. 205–216.
292. M. Springett, M. Mihajlov, E. Brzovska, et al. „An analysis of social interaction between novice older adults when learning gesture-based skills through simple digital games". In: *Univ. Access Inf. Soc.* (Mar. 2021), pp. 1–17.
293. Gunnar Stevens and Volkmar Pipek. „Making use: understanding, studying, and supporting appropriation". In: *Socio-Informatics*. Oxford University Press, 2018.
294. Gunnar Stevens, Markus Rohde, Matthias Korn, et al. „Grounded design. A research paradigm in practice-based computing". In: *V. Wulf; V. Pipek; D. Randall; M. Rohde* (2018), pp. 139–176.
295. Tobias Störzinger, Felix Carros, Anne Wierling, Catrin Misselhorn, and Rainer Wieching. „Categorizing Social Robots with Respect to Dimensions Relevant to Ethical, Social and Legal Implications". In: *i-com* 19.1 (2020), pp. 47–57.
296. Samuel Strauss. „Dashcam und Datenschutz. Eine kritische Gegenueberstellung von alter und neuer Rechtslage". In: *NZV-Neue Zeitschrift fuer Verkehrsrecht* 31.12 (2018).
297. Jörg Strübing. „Was ist Grounded Theory?" In: *Grounded Theory*. Springer, 2004, pp. 13–35.

298. Nicole Strutz, Simone Kuntz, Nils Lahmann, and Anika Steinert. „Analysis of the technical readiness and usage of nursing innovation technologies by personnel in the nursing process“. In: *HeilberufeScience* 11 (2020), pp. 27–34.
299. Kristen Stubbs, Pamela J Hinds, and David Wettergreen. „Autonomy and common ground in human-robot interaction: A field study“. In: *IEEE Intelligent Systems* 22.2 (2007), pp. 42–50.
300. Rachel E. Stuck and Wendy A. Rogers. „Older Adults' Perceptions of Supporting Factors of Trust in a Robot Care Provider“. In: *J. Rob.* 2018 (2018).
301. Mathew J Summers, Innocenzo Rainero, Alessandro E Vercelli, et al. „The My Active and Healthy Aging (My-AHA) ICT platform to detect and prevent frailty in older adults: Randomized control trial design and protocol“. In: *Alzheimer's & Dementia: Translational Research & Clinical Interventions* 4.1 (2018), pp. 252–262.
302. S. Shyam Sundar, T. Franklin Waddell, and Eun Hwa Jung. „The Hollywood robot syndrome: Media effects on older adults' attitudes toward robots and adoption intentions“. In: *Penn State* (2016), pp. 343–350.
303. Elizabeth A Suter. „Focus groups in ethnography of communication: Expanding topics of inquiry beyond participant observation“. In: *The Qualitative Report* 5.1 (2000), pp. 1–14.
304. Ryuichi Tanioka, Rozzano Locsin, Yuko Yasuhara, Tetsuya Tanioka, et al. „Potential legal issues and care implications during care-prevention gymnastic exercises for the elderly using Pepper in long term health care facilities“. In: *Intelligent Control and Automation* 9.03 (2018), p. 85.
305. Nick Taylor, Keith Cheverst, Peter Wright, and Patrick Olivier. „Leaving the wild: lessons from community technology handovers“. In: *Proceedings of the SIGCHI Conference on Human Factors in Computing Systems*. 2013, pp. 1549–1558.
306. Hilda Tellioğlu, Myriam Lewkowicz, Aparecido Fabiano Pinatti De Carvalho, Ivan Brešković, and Marén Schorch. „Collaboration and coordination in the context of informal care (CCCiC 2014)“. In: *CSCW Companion '14: Proceedings of the companion publication of the 17th ACM conference on Computer supported cooperative work & social computing*. New York, NY, USA: Association for Computing Machinery, Feb. 2014, pp. 339–342.
307. „The New Era of Mixed Methods“. In: *Journal of Mixed Methods Research* 1.1 (2007), pp. 3–3. eprint: https://doi.org/10.1177/1558689806293042.
308. Dominic M. Thomas and Robert P. Bostrom. „Team leader strategies for enabling collaboration technology adaptation: team technology knowledge to improve globally distributed systems development work“. In: *European Journal of Information Systems* 19.2 (2010), pp. 223–237.
309. Priyesh Tiwari, Jim Warren, Karen Day, et al. „Feasibility study of a robotic medication assistant for the elderly“. In: *Proceedings of the Twelfth Australasian User Interface Conference-Volume 117*. 2011, pp. 57–66.
310. Hermann Troger. „Demografische Entwicklungen“. In: *7 Erfolgsfaktoren für wirksames Personalmanagement*. Springer, 2016, pp. 3–14.
311. Fekede Tuli. „The basis of distinction between qualitative and quantitative research in social science: Reflection on ontological, epistemological and methodological perspectives“. In: *Ethiopian Journal of Education and Sciences* 6.1 (2010).

312. Kenneth J. Turner and Marilyn R. McGee-Lennon. „Advances in telecare over the past 10 years“. In: *SHTT* 1 (Nov. 2013), pp. 21–34.
313. Turja Tuuli and Parviainen Jaana. „The Use of Affective Care Robots Calls Forth Value-based Consideration“. In: *2020 29th IEEE International Conference on Robot and Human Interactive Communication (RO-MAN)*. 2020, pp. 950–955.
314. Achim Uhl. „Wohnwelten 2025“. In: *BdW Blätter der Wohlfahrtspflege* 161.3 (2014), pp. 109–112.
315. David Unbehaun, Konstantin Aal, Felix Carros, Rainer Wieching, and Volker Wulf. „Creative and Cognitive Activities in Social Assistive Robots and Older Adults: Results from an Exploratory Field Study with Pepper“. In: *Proceedings of the 17th European Conference on Computer-Supported Cooperative Work-Demos and Posters*. 2019.
316. David Unbehaun, Konstantin Aal, Daryoush Daniel Vaziri, et al. „Facilitating collaboration and social experiences with videogames in dementia: results and implications from a participatory design study“. In: *Proceedings of the ACM on Human-Computer Interaction* 2.CSCW (2018), pp. 1–23.
317. David Unbehaun, Daryoush Vaziri, Konstantin Aal, et al. „MobiAssist-ICT-based Training System for People with Dementia and Their Caregivers: Results from a Field Study In Proceedings of the 2018 ACM Conference on Supporting Groupwork (GROUP '18)“. In: *ACM, New York, NY, USA* (2018), pp. 122–126.
318. David Unbehaun, Daryoush Daniel Vaziri, Konstantin Aal, et al. „Exploring the Potential of Exergames to affect the Social and Daily Life of People with Dementia and their Caregivers“. In: *Proceedings of the 2018 chi conference on human factors in computing systems*. 2018, pp. 1–15.
319. Tijs Vandemeulebroucke, Bernadette Dierckx de Casterle\elseé\fi, and Chris Gastmans. „The use of care robots in aged care: A systematic review of argument-based ethics literature“. In: *Arch. Gerontol. Geriatr.* 74 (2018), pp. 15–25.
320. Ivaylo Vassilev, Alison Rowsell, Catherine Pope, et al. „Assessing the implementability of telehealth interventions for self-management support: a realist review“. In: *Implementation Sci.* 10.1 (Dec. 2015), pp. 1–25.
321. Lucie Vidovićová and Tereza Menšíková. „Materiality, Corporeality, and Relationality in Older Human–Robot Interaction (OHRI)“. In: *Societies* 13.1 (2023), p. 15.
322. Stephen Viller. „The group facilitator: a CSCW perspective“. In: *Proceedings of the Second European Conference on Computer-Supported Cooperative Work ECSCW'91*. Springer. 1991, pp. 81–95.
323. Kazuyoshi Wada and Takanori Shibata. „Living with seal robots—its sociopsychological and physiological influences on the elderly at a care house“. In: *IEEE transactions on robotics* 23.5 (2007), pp. 972–980.
324. Alan R Wagner, Jason Borenstein, and Ayanna Howard. „Overtrust in the robotic age“. In: *Communications of the ACM* 61.9 (2018), pp. 22–24.
325. Ina Wagner. „Critical reflections on participation in design“. In: *Socio-Informatics*. Oxford University Press, 2018.
326. James R. Wallace, Saba Oji, and Craig Anslow. „Technologies, Methods, and Values: Changes in Empirical Research at CSCW 1990–2015“. In: *Proc. ACM Hum.-Comput. Interact.* 1.CSCW (Dec. 2017).
327. Lin Wan, Claudia Müller, Volker Wulf, and David William Randall. „Addressing the Subtleties in Dementia Care: Pre-Study and Evaluation of a GPS Monitoring System“.

In: *Proceedings of the SIGCHI Conference on Human Factors in Computing Systems*. CHI '14. New York, NY, USA: Association for Computing Machinery, 2014, pp. 3987–3996.
328. Xi Vincent Wang and Lihui Wang. „A literature survey of the robotic technologies during the COVID-19 pandemic". In: *Journal of Manufacturing Systems* (2021).
329. Philine Warnke, Andreas Röß, and Cornelia Reimoser. *Dokumentation des Workshops Ethik & interne Programme Fraunhofer RRI Roadmaps*. n. d.
330. Philip Weber, Kevin Krings, Julia Nießner, Sabrina Brodesser, and Thomas Ludwig. „FoodChattAR: Exploring the Design Space of Edible Virtual Agents for Human-Food Interaction". In: *Designing Interactive Systems Conference 2021*. DIS '21. Virtual Event, USA: Association for Computing Machinery, 2021, pp. 638–650.
331. Thilo Weichert. „Praktische Anwendungsprobleme im Gesundheitsdatenschutz". In: *Medizinrecht* 37.8 (2019), pp. 622–625.
332. Katharina Weitz, Dominik Schiller, Ruben Schlagowski, Tobias Huber, and Elisabeth André. „"Do you trust me?" Increasing user-trust by integrating virtual agents in explainable AI interaction design". In: *Proceedings of the 19th ACM International Conference on Intelligent Virtual Agents*. 2019, pp. 7–9.
333. William Foote Whyte. *Participant observer: An autobiography*. Cornell University Press, 1994.
334. Vincent Wiegel. „Wendell Wallach and Colin Allen: moral machines: teaching robots right from wrong". In: *Ethics and information technology* 12.4 (2010), pp. 359–361.
335. Marion Wiethoff, Thierry Meulenbroek, Hans Stafleu, and Rogier van Boxtel. „Participative design for home care nursing". In: *Mobile Virtual Work*. Springer, 2006, pp. 203–229.
336. Kenneth W WOODHOUSE, Hilary Wynne, Shelagh Baillie, OFW James, and MD Rawlins. *Who are the frail elderly?* 1988.
337. World Health Organization. *WHO announces COVID-19 outbreak a pandemic*. [Online; accessed 19. Mar. 2021]. Mar. 2020.
338. Ya-Huei Wu, Wrobel, Jeeeeeee\elseé\fie, Souad, and Anne-Sophie Rigaud. „Acceptance of an assistive robot in older adults: a mixed-method study of human-robot interaction over a 1-month period in the Living Lab setting". In: *Clin. Interv. Aging* 9 (2014), pp. 801–811.
339. Volker Wulf. *Anpaßbarbarkeit im Prozeß evolution ä rer Systementwicklung*. 1994.
340. Volker Wulf and Matthias Jarke. „The economics of end-user development". In: *Communications of the ACM* 47.9 (2004), pp. 41–42.
341. Volker Wulf, Claudia Müller, Volkmar Pipek, et al. „Practice-based computing: Empirically grounded conceptualizations derived from design case studies". In: *Designing socially embedded technologies in the real-world*. Springer, 2015, pp. 111–150.
342. Volker Wulf, Volkmar Pipek, David Randall, et al. *Socio-informatics*. Oxford University Press, 2018.
343. Volker Wulf, Markus Rohde, Volkmar Pipek, and Gunnar Stevens. „Engaging with practices: design case studies as a research framework in CSCW". In: *Proceedings of the ACM 2011 conference on Computer supported cooperative work*. 2011, pp. 505–512.
344. Sangseok You, Lionel P. Robert, and Soo Young Rieh. „The Appropriation Paradox: Benefits and Burdens of Appropriating Collaboration Technologies". In: *Proceedings of*

the 33rd Annual ACM Conference Extended Abstracts on Human Factors in Computing Systems*. CHI EA '15. New York, NY, USA: Association for Computing Machinery, 2015, pp. 1741–1746.

345. Shuai Yuan, Simon Coghlan, Reeva Lederman, and Jenny Waycott. „Social Robots in Aged Care: Care Staff Experiences and Perspectives on Robot Benefits and Challenges". In: *Proceedings of the ACM on Human-Computer Interaction* 6.CSCW2 (2022), pp. 1–23.
346. Ye Yuan and Svetlana Yarosh. „Beyond Tutoring: Opportunities for Intergenerational Mentorship at a Community Level". In: *CHI '19: Proceedings of the 2019 CHI Conference on Human Factors in Computing Systems*. New York, NY, USA: Association for Computing Machinery, 2019, pp. 1–14.
347. Hsiu-Ping Yueh, Weijane Lin, S-Chen Wang, and Li-Chen Fu. „Reading with robot and human companions in library literacy activities: A comparison study". In: *British Journal of Educational Technology* 51.5 (2020), pp. 1884–1900.
348. Setareh Zafari, Isabel Schwaninger, Matthias Hirschmanner, et al. „"You Are Doing so Great!"—The Effect of a Robot's Interaction Style on Self-Efficacy in HRI". In: *2019 28th IEEE International Conference on Robot and Human Interactive Communication (RO-MAN)*. IEEE, Oct. 2019, pp. 1–7.
349. Irati Zamalloa, Risto Kojcev, Alejandro Hernández, et al. „Dissecting Robotics-historical overview and future perspectives". In: arXiv preprint arXiv:1704.08617 (2017).
350. Jakub Złotowski, Kumar Yogeeswaran, and Christoph Bartneck. „Can we control it? Autonomous robots threaten human identity, uniqueness, safety, and resources". In: *International Journal of Human-Computer Studies* 100 (2017), pp. 48–54.
351. Jan C. Zöllick, Adelheid Kuhlmey, Ralf Suhr, et al. „Akzeptanz von Technikeinsatz in der Pflege". In: *Pflege-Report 2019: Mehr Personal in der Langzeitpflege—aber woher?* Ed. by Klaus Jacobs, Adelheid Kuhlmey, Stefan Greß, Jürgen Klauber, and Antje Schwinger. Berlin, Heidelberg: Springer Berlin Heidelberg, 2020, pp. 211–218.

The manufacturer's authorised representative in the EU is Springer Nature Customer Service Centre GmbH, Europaplatz 3, 69115 Heidelberg, Germany. If you have any concerns regarding our products, please contact ProductSafety@springernature.com

Printed and bound by CPI Group (UK) Ltd, Croydon, CR0 4YY
30/09/2026
02225566-0003